INTRODUCTORY CHEMISTRY PROBLEMS

The Chemistry and Physics Behind the Numbers

INTRODUCTORY CHEMISTRY PROBLEMS

The Chemistry and Physics Behind the Numbers

Michael E Green

City University of New York, USA

NEW JERSEY · LONDON · SINGAPORE · BEIJING · SHANGHAI · HONG KONG · TAIPEI · CHENNAI · TOKYO

Published by

World Scientific Publishing Co. Pte. Ltd.

5 Toh Tuck Link, Singapore 596224

USA office: 27 Warren Street, Suite 401-402, Hackensack, NJ 07601

UK office: 57 Shelton Street, Covent Garden, London WC2H 9HE

British Library Cataloguing-in-Publication Data
A catalogue record for this book is available from the British Library.

INTRODUCTORY CHEMISTRY PROBLEMS
The Chemistry and Physics Behind the Numbers

Copyright © 2026 by World Scientific Publishing Co. Pte. Ltd.

All rights reserved. This book, or parts thereof, may not be reproduced in any form or by any means, electronic or mechanical, including photocopying, recording or any information storage and retrieval system now known or to be invented, without written permission from the publisher.

For photocopying of material in this volume, please pay a copying fee through the Copyright Clearance Center, Inc., 222 Rosewood Drive, Danvers, MA 01923, USA. In this case permission to photocopy is not required from the publisher.

ISBN 978-981-98-0830-4 (hardcover)
ISBN 978-981-98-0932-5 (paperback)
ISBN 978-981-98-0831-1 (ebook for institutions)
ISBN 978-981-98-0832-8 (ebook for individuals)

For any available supplementary material, please visit
https://www.worldscientific.com/worldscibooks/10.1142/14186#t=suppl

Typeset by Stallion Press
Email: enquiries@stallionpress.com

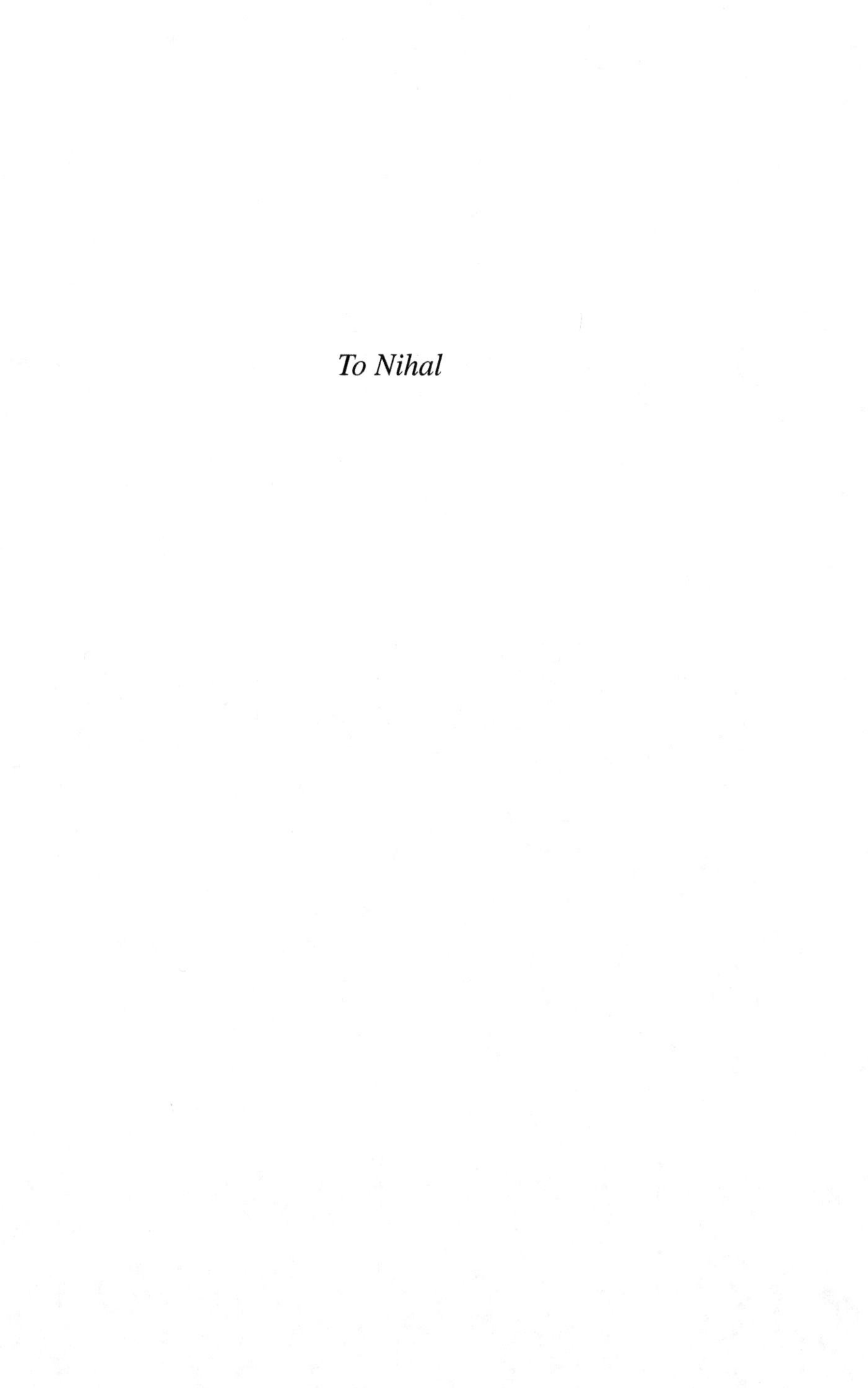

To Nihal

Acknowledgements

This book is the product of over half a century of teaching chemistry at City College of New York. I have to thank many colleagues over the years, as well as an atmosphere that made teaching a central part of one's responsibilities as a faculty member.

I attempted to put together a collection of problems something like this book about 30 years ago, with Denise Garland, whom I must also thank.

This book owes special thanks to a colleague, Virinder Parma, who contributed comments on the final version.

Finally, I would like to thank Joy Quek for her able assistance with the manuscript.

Preface

A Note From the Author

Many years ago, I decided to take drawing lessons, for which I had no background at all. It was a beginner's class, and all the students, including me, made mistakes — something like this:

Preface Fig. 1. Drawing by a beginner

The head is way too big. A leg shrinks to fit in the paper. The drawing does not represent a possible human being. The teacher would look at this kind of mistake and say "Oh come now". He would point out the obviously ridiculous proportions. In a drawing class you would have to figure out what to do about it. Comparable errors in a chemistry class would be likely to draw more specific comments from a teacher; still, as a student, you would have to begin to think about how to approach such problems.

In the drawing class, after about one year the drawings converged to something like preface Fig. 2.

At least the proportions in Fig. 2 are something like those of a human being — still not the drawing skill of an old master, but not ridiculous. This is the biggest step — the second drawing is not good enough for an exhibition, but it is a basis for progress to a more advanced level.

Preface Fig. 2.   After about one year of class, the figure is still not very good, but is plausibly human

In decades of teaching chemistry, I would sometimes see answers on exam questions that made me want to say "Oh come now". This book is the way I am saying "Oh come now" — and hoping that the examples here help in moving from the state of "Oh come now" answers to a reasonably competent state, in which you can understand what is really going on with the chemistry that is described by the conditions in the problem — not a bunch of symbols that you have to memorize, but instead formulas that express what is happening inside a system that is undergoing a chemical change. If a problem concerns a chemical reaction in which you start with 10 grams of material, and your answer says the reaction ends with 34 grams, this is ridiculous. Matter can neither be created nor destroyed; you can't wind up with more than three times (even 1.01 times) as much material as you started with. Getting an amount of matter at the end of a reaction different from what you had to start is like Preface Figure 1; it is a ridiculous result. There are other ridiculous answers in different kinds of problems. You can't have an atom as big as an elephant, or even a bacterium. These kinds of errors can result from remembering a formula incorrectly, or even from pushing the wrong button on a calculator. Neither of these is unusual or fatal; leaving the results without thinking about what the numbers mean is fatal — the latter means you don't understand what you are doing. I hope this book helps you in seeing what it means to understand what the problem really is describing, and finding a path to reach this understanding. At the end of the first year of chemistry, you should be able to do as well as Preface Figure Two — not showing mastery, but making a basis for progress. This book should help you get to the point at which you look for the physical and chemical meaning of the value you find when you solve a problem, and thus help you decide whether your answer is reasonable or ridiculous.

My drawing and painting have now progressed to the point where I can do oil paintings that are definitely not masterpieces, but are good enough that I can reasonably save them, and feel good about having produced them. Hopefully, once you get to thinking about chemistry as a set of processes happening in the real world, you will be able to go on to more advanced courses, in which you can no longer even begin to get by just with memorized equations. You can feel good about having progressed to understanding chemistry at a level that is at least useful, and maybe more.

In introductory chemistry courses, a fairly large fraction of the students fail to make it past the exams, as they cannot do the problems. Often, these students have not attempted to learn what the problems are about — they have not tried to learn the underlying chemistry. Instead, they memorize formulas, and then plug in data. After a while, the formulas get too complex, and too numerous, for them to be memorized, and the student is lost. Some students work very hard their first year, and are really good at memorizing, so they survive the first year, only to find themselves hopelessly lost by the second year, when the amount of material that would have to be memorized becomes overwhelming. If a student has not understood why the formulas work, where they come from and how they can be rearranged so as to give different views of the chemical system, the student cannot proceed.

Common errors involve transposing symbols so that something that should have been multiplied is divided, or vice versa. An extreme, and fortunately rare case, comes when a student is asked to find the mass of a hydrogen atom, inverts the formula, multiplying the atomic mass by Avogadro's number instead of dividing, and gets about one-tenth the mass of the moon. The moon is big; the hydrogen atom is small. Here there is an error of about 47 orders of magnitude, so large just about anyone can notice it. The error would be avoided if the student simply paid attention to the *meaning* of the numbers. A case that actually occurred on an exam: a student was given a temperature and pressure of a diatomic gas, and asked to find the pressure at the same temperature if the gas partially dissociated. The pressure was proportional to the total number of molecules, dissociated plus undissociated. The initial pressure was 0.1 atm. The student got 707 atm and complained bitterly when he lost more than half credit. He had certainly started from the right equations, and essentially followed the complete procedure. Of course, if he had stopped to think, he would have realized that if all the gas had dissociated, the pressure could increase to 0.2 atm, the absolute upper limit for the conditions of the problem. Leaving 707 atm meant that he had not understood what he was doing, and deserved to lose almost all credit. If he had simply said that there was an arithmetic error, and the maximum pressure was 0.2 atm, he might have gotten almost full credit. He lost credit because he showed that he did not understand what the problem was about. His best argument was that most students did not stop to think about what the numbers meant. Unfortunately, this is true. It is the aim of this book to help make this no longer true.

Standard chemistry texts all give worked out problems, but these rarely sufficiently emphasize the importance of understanding what is actually possible in a real physical system with whatever conditions are given in the problem. The student is not asked to consider whether an answer is plausible, and why, nor what would make the answer out of bounds. There are limits on what can be possible, as in the gas dissociation problem above. If the student had been told to first estimate an upper and lower bound to the answer, in this case 0.2 and 0.1 atm, and say why these were the limits, the ridiculous answer would not have appeared on the exam paper.

Sometimes, of course, standard texts do use approximations; they are almost inevitable in certain equilibrium problems, and some weak acid dissociation problems. As soon as you begin to learn about acids and bases, you know immediately that if there is a large concentration of H^+ there is a small concentration of OH^-. However, this approach is rarely carried over to other problems, nor is there any attempt to make this part of a general approach to problem solving, at least in most standard texts. Even in the weak acid dissociation problems, it is not made clear that the approximation cannot be used if the concentration is less than or equal to the dissociation constant of the acid. In this book we attempt to make this part of a general approach to problem solving. A number of problems are given for each of the quantitative aspects of introductory chemistry, and for each problem, the solution is preceded by an estimate of the answer — we usually settle for the order of magnitude, or, if this is immediately obvious, perhaps a one significant figure answer. We may look for upper and lower bounds, when these are appropriate. In each case, the estimate is followed by the conventional solution, so that the student can see how the estimate relates to the final answer. Problems are chosen to illustrate the main physical and chemical principles that underlie that type of problem, and to the chapter topic. The number of problems is small, but each comes with some discussion. I hope that this approach will help the student get past the memorization of formulas, and help him or her to think like a chemist.

A final note: In a few places, some redundancy has been left in to avoid the inconvenience of having to look back for something needed in a later place.

Contents

Acknowledgements — vii
Preface — ix

1 "Reasonable Answers?" — 1
2 Some Convenient Arithmetic Operations — 9
3 Part One: Dimensional Analysis and Units — 17
3 Part Two: The Standard International (SI) System of Units — 23
4 Atomic Mass and Molecular Mass — 33
5 Gases — 41
6 Stoichiometry — 55
7 Atomic Structure and Spectra — 69
8 Solids and Unit Cells — 97
9 Solution Calculations — 113
10 Chemical Equilibrium with Equilibrium Constants K_{eq}, K_c, K_p — 133
11 Thermochemistry: Basic Ideas of Thermodynamics; The Background of the Calculations on Equilibrium — 147
12 Solubility — 169
13 Acids, Bases, and pH — 181
14 Chemical Kinetics — 195

15	Electrochemistry	213
16	Nuclear Chemistry	227
Epilogue		235
Appendix: An Extra Problem, with Some Extra Complications		237
Index		245

1

"Reasonable Answers?"

In the preface, we talked about the kind of ludicrous errors that were possible if you don't think about what you are doing, and try to get by just memorizing equations. After a little while the amount of stuff to memorize becomes overwhelming, and you are lost. This book tries to get you to think about what you are doing, so that you are able to use the equations, and know when you have done something wrong, especially if it is very wrong. Very often, if you do make a mistake, it is a very big mistake, and you should be able to tell right away that the answer is not possible. If you do this kind of check, you may not always get the right answer, but you won't do anything ridiculous, and more often than not, you will catch the mistake and be able to correct it.

Memorizing equations is possible for a computer. You are not a computer. Don't try to be one. Computers don't know what they are doing (at least not so far — artificial general intelligence is threatening to change that, but it isn't there yet). You, on the other hand, have to know what you are doing, or you will make errors that will be worse and worse as the material becomes more complex. If you want to go on to courses that use the first year course in chemistry, whether more advanced chemistry, or other science courses in biology, or physics (where similar considerations apply), memorized equations will do no good. However, if you learn to think in quantitative terms like a scientist, whether in chemistry, physics, or in biology as well, you can move forward and learn the more advanced material just as easily as you learned

1

the elementary material. The advances in understanding biology over the past three decades have been immense, and there is now much more emphasis on understanding, rather than remembering categories of things; before the structure and sequence of DNA and proteins could be determined routinely, fundamental understanding, for the most part, was qualitative and conceptual.

In working out estimates of problems, you will be able to develop a kind of chemical intuition. Once you have some feel for what is reasonable, it becomes almost impossible to produce a silly answer. Sometimes you plug numbers into a calculator, using a memorized equation, and take whatever shows up on the calculator display as THE ANSWER. As Sportin' Life sang in *Porgy and Bess,* "It Ain't Necessarily So". You should realize that the answer must have a relation to reality. Sometimes an adequate estimate amounts to doing the problem to one significant figure (SF) without a calculator (one SF means rounding everything to its first digit that is not a zero, and keeping track of the decimal point — we will talk about this more later). For this, you don't need a calculator. A more sophisticated estimate comes from noticing that some quantity sets an upper limit, or lower limit, on the answer. If you have a chemical reaction, in which you start with ten grams of reactants, you better finish with ten grams of reactant plus product. If your answer is 244 grams, or anything but ten grams, something is very wrong. Unfortunately, some students do wind up with such a ridiculous answer. Thinking about what you are doing is enough for you to at least write down an OOPS on the exam, if that is the answer you get, so the grader can realize that you know what you are doing, even if you did it wrong. In other words, you can develop a kind of "common sense" for chemistry, just as for all the other things in life. You don't have it yet, because you have no experience with chemistry; this book is intended to help you develop this kind of common sense.

The problems are set at several levels: sometimes we begin with a completely trivial example, something that just illustrates the point that we are trying to make. These will be followed by examples that require more effort, but still depend on what was illustrated by the initial trivial example. This should enable you to see how the complication develops, and in what sense the problems really show the same point; the more difficult problem extends the elementary ideas that were first introduced in the simple example. Elementary here means the elements of which the problem is compounded, that is, the elementary ideas, which can be put together with others to allow the solution of problems that are realistic, or at least more interesting. We also include problems that really are difficult, and force you to think about what is happening in the chemical system. We invite you to extend the problems, using your own intuition, to new systems — this will help you develop your understanding of what the ideas in the problem really allow. In many problems, we take account of the limits of the accuracy of the data presented; there is no explicit statement that you should extend the problems, but you will learn more if you do. If the error in some input quantity is, say, 3%, and the output is proportional to this quantity, does it make sense to worry whether another quantity is accurate to 0.1% or 0.01%? Unless that quantity is perhaps in an exponent, or raised to a very high power, probably not. There will always be at least that 3% error in the final result. Again, a kind of chemical (or, more generally, scientific) common sense will tell you not to bother with all the effort of redoing the calculation when this other quantity improves from plus or minus 0.1% to 0.01%.

No matter what the problem, finding the estimated answer is not necessarily easier than the final calculation; for the estimate you have to think about what is involved in getting to the final answer. To get the final answer, you may then just have to use your calculator. What you will find is that after you do a number of estimates the problems and the solutions begin to make sense, in a way that no number of repetitions of algorithmic solutions would allow. Only a relatively small number of examples are given in this book. If you do these, we hope that you apply the reasoning whenever you do problems.

There are some classes of problems that absolutely require an estimate to begin with, especially if an extended set of simultaneous equations are required. This happens with some chemical equilibria calculations. Often, only by noticing which quantities are large, and which are small enough to be neglected, is it possible to do the problem at all. This means that you have to have a sense of the order of magnitude of the plausible range of possible results. Sometimes, this is straightforward; if you are in acid solution, the amount of base (as you will learn later, this means the concentration of hydroxide ion) is small, and can generally be neglected if added to anything present in more normal concentrations. For example, you may have a problem with several equations, not all simple or linear. If, however, an equation adds a quantity that is 10,000 times smaller than the quantity to which it is being added, it is probably reasonable to simply remove that quantity from the equation. Then, with one less variable, the equation may become much easier to solve. There is also a procedure if you need an extremely accurate answer that requires the small quantity to be included, but that is almost always a more advanced topic.

Computers show up in practically everything now, and AI is beginning to encroach on more and more parts of life. However, if you use a computer to solve a problem, you still need to be sure that what you did makes sense. Did you enter the data correctly? Assuming you are using a debugged, commercial, program suite, is the software designed to calculate the range of values you are calculating? Sometimes software assumes that a quantity is so much smaller than another that it can be neglected, as we suggested in the last paragraph. Have you checked that this so in your problem? We will not discuss much having to do with computers in this book, as it is intended to develop your intuition, not that of an artificial intelligence, but if you do depend on artificial intelligence, beware of everything from how you asked the question to possible hallucinations on the part of the AI. It is still worth knowing what you are doing. In fact, that is the theme of this entire book: understand what you are doing, so don't try to imitate a computer.

There are two major techniques that we will return to throughout the book: Limits, or upper and lower bounds, and orders of magnitude. We also pay careful attention to the accuracy of the answer, in relation to the data given in the problem. If the original data are plus or minus 1%, we should not report an answer seemingly accurate to 0.1%. Calculators may give eight figure "accuracy" (error only in the eighth place, or one millionth of one percent). Nature rarely gives such accuracy.

1) Order of magnitude calculations: the estimated answer gives a single value, usually to only one significant figure. These estimates can be made by rounding all quantities to either the first figure, but keeping track of the decimal point, or, even more roughly, just to the nearest power of ten. This can give a quick estimate before beginning (although just using the nearest power of ten may, if several sequential calculations are required, lead to an error of more than a factor of ten — errors accumulate, so if a value is off by 10% in step one, it may become much more if you look at a calculation with, say, five steps, especially if the number appears in an exponent in any part of the problem). The same kind of rough estimate may be used to check that the final answer in a problem is not unreasonable. If you use the first "significant figure" (we will define this later, but here just take it to mean the first digit in either the data or the answer), this may be more useful. For example, suppose you wanted to find the volume of the pit

left by a strip mining operation. The dimensions are 363 m × 218 m × 57 m. Now obviously the quickest thing to do is just plug these numbers into a calculator. However, let's look at the two techniques, order of magnitude and one significant figure, we just introduced: for order of magnitude only, take $100 \times 100 \times 100 = 10^6$ m^3. This is good enough to prevent us from utterly ridiculous results, but even a bit of quick mental arithmetic would be better. If we use the first significant figure we have: $400 \times 200 \times 60 \approx 5 \times 10^6$ m^3. This is better, and now we should go ahead and use our calculator to get 4.51×10^6 m^3. Our estimate tells us that this is reasonable. We should not carry the calculation further, as the input numbers are of only the accuracy shown. In fact, following the rules of significant figures which we will define later, the answer is 4.5×10^6 m^3, and the final 1 after the 5 is not justified; if each of the dimensions is good to ±1 m, by the time you multiply the three numbers, the uncertainty is more than ±1 m^3. While this is a particularly trivial example, the principle illustrated will be carried through the entire book. The problem with the order of magnitude estimates is that you are really just doing the problem following the algorithm, and it does not really require that you understand what you are doing. It does usually avoid ridiculous errors; these often come in when we plug in and make a mistake in arithmetic, which even this simple level of estimate avoids.

2) The setting of upper and lower bounds for the answer is a little more sophisticated. The example we just did is almost too trivial to apply this, although if we observe that in rounding 363 and 218 to 100, we made a more drastic approximation than in rounding 57 to 100, so 10^6 m^3 is a lower bound. Let's look at a more interesting, even if still trivial, case: suppose 2 g of hydrogen (H) react with 10g oxygen(O). Obviously, we can't possibly get more than 12 g of product. This is an immediate upper bound. Can we set a lower bound? With what we

know at this point (if you have not gotten to moles, and stoichiometry), not so easily. We don't know whether all the oxygen reacts, with some left over hydrogen, or all the hydrogen reacts and there is some left over oxygen, or the amounts are exactly matched, and nothing is left over. If all the hydrogen reacts we have more than two grams of product (more, because at least some oxygen must be in the product). If all the oxygen reacts we have at least 10 grams of product. The best we can do for a lower bound is therefore >2 g. Of course, once we know a little about how to do such problems in the course, we can do much better. Even with this little bit of effort, we see that we have to have some sense of what we are doing, including asking ourselves what it is that we are looking for, and what chemical possibilities exist — left over hydrogen, left over oxygen, or just matched to have nothing left over. Then the procedure from that point is obvious in this problem. It takes more effort with more elaborate problems, but the basic idea is not changed.

Summary: Overall, the book will attempt to make it possible to develop some chemical intuition, and a sense of what makes sense, or doesn't. There are several procedures that can be applied, and we will introduce the relevant ones. We should be able to avoid having answers more than ten times too big, or ten times too small, and avoid claiming impossible accuracy for calculations with not-so-accurate data.

2

Some Convenient
Arithmetic Operations

Let us assume that, if you are taking chemistry, you know how to solve a quadratic equation, and how to do elementary operations with logarithms, both natural and base 10. Just to make sure, recall that $\ln x = 2.303 \log_{10} x$, that $\ln (a/b) = \ln (a) - \ln (b)$, and other fundamental relations. Two quantities, say x and y, are directly proportional if $y = ax + b$, and this is called a linear equation. If $xy = c$, with c a constant, then x and y are inversely related, and this is a non-linear relation. Because linear relations are much easier to deal with, it may be convenient in solving a problem to define a variable $z = 1/y$, and then $x = cz$, which again is linear; however, effort may be required to get the reciprocal variable.

If you have a non-linear relation of the form $y = x^N$, then another way to make it linear is to use the log: Define $z = \ln y$, and $w = \ln x$, so $z = N w$, and the logs at least form a linear relation. Sometimes linear forms are so much easier to solve that it is worth the effort to make this transformation.

Square roots: There are some arithmetical tricks that are worth knowing, starting with taking square roots. If you need the square root of a number that is a square, say a^2, plus something smaller, b^2, it can be approximated as follows:

$$(a^2 + b^2)^{1/2} \approx a + b^2/2a$$

Since we know that a^2 is perfect square, it is not a problem to find a from a^2. We don't have to take the square root of b^2, so we have a good approximation of the overall square root without the effort of actually working out the root. If we square $a + b^2/2a$, we get $a^2 + b^2 + b^4/4a^2$. The last term is the error from the approximation, and if $b < a$, it is, depending on the accuracy needed, very likely negligible. The worst case is $2^{1/2} = (1^2 + 1^2)^{1/2} = 1.414$; $a^2 = b^2 = 1$, and the approximation gives 1.5, about 6% error. If the numbers are further from equal, say $10^{1/2} = (9 + 1)^{1/2} \approx 3\ 1/6 = 3.1667$, compared to 3.1622, good enough to three places. Since everyone has a calculator, this is not important for numerical square roots, but it does become important if you have two quantities whose numerical values are not known, but their relative magnitudes are known. It is useful, for example, in solving quadratic equations, in which the radical $(b^2 - 4ac)^{1/2}$ is required. If $b^2 > 4ac$, then the approximation gives $b - 4ac/2b$. Of course, ac may be negative, in which case $-4ac$ is positive. If $4ac$ is positive and greater than b^2, the root is imaginary, and the solution to the quadratic equation is a complex number; no approximation can make this real.

Essentially, this is an example of a binomial expansion of the square root. Just as we can expand

$$(a + b)^2 = a^2 + 2ab + b^2,$$

we can expand the square root, and the approximation above is the consequence.

Extending this to other roots: We can similarly expand $(a + b)^n = a^n + n\,a^{n-1}\,b + n(n-1)\,a^{n-2}b^2 + \ldots$

And $(a^n + b^n)^{1/n} = a + b^n/na^{n-1} + \ldots$

For $26^{1/3} = (27 - 1)^{1/3} = 3 + (-1/3 \times 9) = 3 - 1/27 = 2.9630$, compared to the correct value $2.9625\ldots$

Again, to be fair, I should say that I used a calculator to get the correct value. For numerical cases, a calculator is perfectly fair. However, once again, we sometimes have cases in which we need an actual algebraic form, and then the expansion is useful. For numerical values, using logs, with the relation $\log (a^n) = n \log a$, for either natural logs or base 10 logs, is often a convenient way to get the answer. Given a quantity Q for which the cube root, for example, is needed, take $(1/3)\log Q = \log Q^{1/3}$, and then take the antilog of $(1/3)\log Q$. In addition, consider the trick we used to linearize the inverse relation, in which we defined a new variable $z = 1/y$; similarly, here we could look for a new variable that simplifies the expression in any way that is convenient.

More on expansions: Powers and roots are not the only functions that can be expanded to get a reasonable or solvable problem. In general, expansions, in order to be useful, require that one quantity be small enough, either compared with 1, or with some other quantity in the expression. Useful expansions include:

$\ln (1 + x) \approx x$ $(x \ll 1)$; note that if you have two terms, a and x, and need $\ln(a + x)$, $x \ll a$, you can use this relation by dividing the argument by a: $\ln(a(1 + x/a)) = \ln(a) + \ln (1 + x/a) \approx \ln(a) + x/a$.

$\exp(x) \approx 1 + x$ $(x \ll 1)$; this is sometimes important in the expression $\exp(-E/RT)$, where RT is thermal energy and E is some other energy; $\exp(-E/RT)$, occurs often in chemical equilibrium and rate problems. Although $E/RT \ll 1$ is rarely true in real physical situations, it can give a useful limiting law at low energy. A century ago, it gave the first reasonable description of dilute ionic solutions.

$\mathrm{Sin} (x) \approx \mathrm{Tan} (x) \approx x$, where x is in radians; a circle is 2π radians $= 360°$. 1 radian $= 57.3°$, and here x must be small compared to 1 radian; similarly $\cos(x) \approx \cot(x) \approx 1 - x^2/2$, with the same condition.

A more advanced trick: If a non-linear expression can be linearized, and thus made solvable, by dropping a term, or approximating it by something linear, then the solution to the linearized form is a starting point. The missing part can then be assumed to be small (let's hope!), and a correction added. Perhaps this extra term leads to a linearized form when thus approximated, or is otherwise calculable. Sometimes, as happens in quantum mechanics, it makes sense to have first order, and then second or even higher order corrections. Such an expansion is called a *perturbation expansion*. While you will not encounter this in first year chemistry, it is worth bearing in mind that there are approximate methods of dealing with problems, and the perturbation expansion is one of the most useful.

SIGNIFICANT FIGURES: In reporting data, or results, it is important to know how much you know, and what your error limits are. One way to express this is by the use of significant figures. Before calculators, there were slide rules, which could be read to three figures. If you wanted better accuracy you could use a big, slow, mechanical calculator which might take a minute to do a multiplication or division, or you could do it by hand. Three figures, 0.1%, was therefore pretty much the maximum available accuracy when slide rules ruled. Now, calculators routinely show eight figures. It may be that three mean something and the last five are just a set of random digits. Here, we want to keep track of which digits we can trust. There are a set of rules that give the main points. You can see that in a many step calculation the errors would accumulate, so that the accuracy would become less than the rules allow. While there are better ways to report accuracy, the significant figure rules do help in becoming acquainted with what we can say, and what we must avoid saying, as the result of a calculation, and they are simple and convenient. Sometimes we have to use some common sense in deciding how many significant figures to report; we can get a tighter error estimate by assuming the last figure is good to half the difference with the value (e.g., 33.1 is between 33.05 and 33.15). By using the values that will produce the largest possible result and the smallest possible result we would have a better estimate of the accuracy with which we know the answer. However, for most of what we need in this book, following the significant figure rules below is good enough. We will use the abbreviation **SF** for significant figures.

1: Defining significant figures. These are figures that actually have information that has meaning, but not the decimal point location. 125 has three SF; so do 0.000125 or 12500000. The zeros indicate the location of the decimal point, but tell us nothing about the accuracy with which we know the number, which is the point of paying attention to SF. Suppose we know the value of 125 to five figures; then write 125.00. There is no reason for the last two zeros, other than to indicate how well we know the number. Another way to indicate any of these numbers is to use standard exponential notation. Thus, 125 becomes 1.25×10^2, 125.00 becomes 1.2500×10^2, 0.000125 becomes 1.25×10^{-4}, and 125000 becomes 1.25×10^5. In this notation, the number of SF is immediately evident.

Some quantities are exact and have an infinite number of SF: the 2 in the expression for *circle circumference* $= 2\pi r$ is exact, and the π has as many SF as we need. We know both these numbers with no error at all.

2: The rules for manipulating these numbers are as follows:
   i) *Multiplying by an exact number* does not affect the number of SF. If we have measured the radius of a circle to be 1.25×10^3 (units), the diameter of that circle is 2.50×10^3 (units), still three SF. The same rule holds for division.
   ii) *Multiplying or dividing* two numbers with finite numbers of SF gives a result with the smaller number of SF. A rectangle with one side measured to be 1.250 (units),

the other 2.05 (units) has an area of 2.56 (units squared), to three SF, because 2.05 limits the number of SF. Note that the units do not affect the number of SFs. It doesn't matter if the lengths are in cm, or inches, or miles, or cubits — the units just have to be all the same.

iii) *Addition and Subtraction*: The last place to have a SF determines the SFs of the sum. For example,

$$98.03$$
$$\underline{+\ 6.2474}$$
$$104.28$$

The last SF is the second decimal place, and we round off. The 74 at the end of 6.2474 is added to nothing in particular, so we have no idea what those places hold. Addition doesn't necessarily do anything terrible to the number of SF, but subtraction of two numbers of similar magnitude leaves us with a very limited knowledge of their difference. For example,

$$98.6359$$
$$\underline{-94.4\ \ \ }$$
$$4.2$$

Here we subtract a three SF number from a six SF number, and the answer has only two SF; the reason is obvious. Beyond that second figure, there is no information. 94.4 is only known to the nearest tenth, so we know nothing about the hundredths place, or anything beyond that. Something like this will always happen in taking the small difference between two large numbers.

iv) *Logarithms*: The number of SF equals the number of SF in the mantissa (the part after the decimal point in the logarithm — before the decimal point indicates the position of the decimal point in the number whose logarithm is shown.

For example: $\log_{10}20.00 = 1.3010$. The mantissa, 3010, has four places, like 20.00. For $\log_{10}20.0 = 1.301$, three places like 20.0. We have $\log_{10}20.005 = 1.3011$, for example.

v) *Many step calculations*: With these, one could follow the same rules for each step, but this is likely to leave a lot of error, as round-off error piles up. Sometimes one could try to carry an extra place, and round off at the end; even though this is still not really legitimate, it can avoid loss of a figure that is still significant — but remember to round off at the end. It is better to keep track of error explicitly to know how accurately the result is really known. Carrying an extra figure is a kind of cheating, claiming more accuracy than is realistic, while losing a legitimate figure throws away hard earned information. Generally, SF rules are only useful for a single step. For serious estimates of error, better methods are available.

QUESTION: Do the following calculation to the correct number of SF:

$$(583 \times 0.088)/(27{,}653 \times 0.22)$$

Estimated answer: Right away, we see two numbers with just two SF, 0.088 and 0.22, so the answer can have only two SF. We can look at the numbers and realize quickly that 0.22 is roughly 1/5, and thus the denominator is around 5000, while the numerator is around 500×0.1, or about 50, so we make a quick estimate of 50/5000, or around 0.01.

Answer: the actual result is 0.0084 to two SF, close enough to the 0.01 we got by just looking at the numbers.

QUESTION: $4 \times 10^6 + 3 \times 10^4 = ?$

Estimated answer: With only one SF, only the 4×10^6 matters, and we neglect the 3×10^4. This is therefore also the final answer. However, if we had had 4.03×10^6 for the first quantity, then the exact answer would be 4.06×10^6 as there would then be enough SF that we could use the second term. Of course, if we had 4×10^6 and 3.03×10^4 we would again wind up with 4×10^6 as the extra SF in the smaller number would be lost compared to the larger number.

$$3$$

Part One: Dimensional Analysis and Units

We divide this chapter into two parts that overlap: dimensional analysis and units, the latter introducing standard international (SI) units. In the first part we look at general principles, in part with a set of antiquated units that illustrate how units can be manipulated, as long as they are consistent, even if not systematic. Then we move to the consistent system of units that we will actually use throughout this book, and everywhere called SI units, in chemistry and physics (and pretty much everywhere in the entire world, outside the U.S.). However, doing a couple of problems with the more difficult antiquated units is instructive, so that is where we begin.

The first point we emphasize is the importance of consistency of the units. You cannot equate feet to pounds, or meters to kilograms. If you add 7 kg to 34 meters you end up with 1 nonsense. You can cancel units in multiplication and division just as you can cancel numbers. For example: If the density of mercury is 13.6 g cm^{-3}, what volume, V, of mercury holds 20.0 g of mercury?

V = 20.0 g/13.6 g cm^{-3} = 1.47 cm^3 to three SF. If the problem had stated simply 20 g, in principle it has one SF, but usually we would take this to mean 20 to two SF, and round off to 1.5 cm^3. Since we were explicitly given three SF, we get 1.47 cm^3. We had cm^{-3} in the denominator, so we bring it up as cm^3. In other words, we treat the unit just as we would a number. The mass unit, g, cancelled in the numerator and denominator: use the same reasoning in the next problem.

Extend the problem: suppose you want the volume of 20.0 mg of mercury. Now the mass unit does not cancel: 1000 mg = 1 g. Formally we could write

$$V = (1 \text{ g}/1000 \text{ mg}) \times 20.0 \text{ mg} \times 13.6 \text{ g cm}^{-3} = 1.47 \times 10^{-3} \text{ cm}^3$$

Of course, in real life, if we had done the first problem, we would simply observe that we have 10^{-3} as much mass, all else being the same, and write down the factor of 10^{-3}. However, by writing it out formally, we see that what we have done is to cancel the mg unit, and returned to g. There are lots of cases in which we make similar conversions, say minutes or hours to seconds (the SI unit of time is seconds), so we get factors like 1 min/60 s, and use this, or its inverse, to convert between minutes and seconds. Note that this conversion factor is exact, having an infinite number of SF. This is valid for all cases. Sometimes you get a sort of local conversion factor — say it is a silver mine, and it is found that a ton of ore has 34 lbs of silver. Using 34 lbs/1 ton of ore is a valid factor for this particular case, but has no general validity.

Problem 2: How many meters is it from the orbit of earth to the orbit of Jupiter, if you approximate the orbits as circular? Given: Earth orbit radius = 93,000,000 miles. Jupiter's orbit is 5.2 times that of earth. 1608 m = 1mile

Here there are a couple of conversions to go through. We need to get from miles to meters, and we need the distance from earth orbit to Jupiter's orbit. The latter is easily found in miles, as we need to simply multiply by 5.2. All the numbers are given to two SF, except the conversion of meters to miles, and we might as well round that off to two figures as well.

Estimate: We can estimate an answer quickly: If earth orbit is rounded to 10^8 miles, and then Jupiter to 5×10^8 miles, we get (round numbers, as in an estimate) 4×10^8 miles. Convert miles to meters: 1.6×10^3 m mi^{-1} $\times 4 \times 10^8$ mi $= 6.4 \times 10^{11}$ m. Note that we have canceled the miles (mi) units. We have a sort of pseudo 2 SF answer, but for the actual answer, we should go back and put in the two SF numbers given in the problem:

Answer: 7.7×10^{11} m. This example is almost too trivial to bother writing it all out. We only had two numbers that we rounded off: the earth orbit was rounded up about 7%, and the 5.2 was rounded down about 4% (but then multiplied by a factor >1) so we should expect the answer to be a little larger than the estimate, and it is. It is worth paying attention to what we do to the numbers.

NOTE: Suppose we had been given 4 SF for the earth's orbit and the distance to Jupiter's orbit. What would the answer look like then? If we are not careful, it would look like nonsense. The orbits are ellipses, not circles, with the planets at one focus of the ellipse. If the error were in the fourth SF, around 10^5 mi, we would be pretty nearly at the diameter of the planet Jupiter. Given the elliptical nature of the orbit, and the size of the planet, it is hard to see that the problem would have any meaning. If you were going to send a spacecraft from earth to Jupiter, you would want to know positions of the planets at the specific times the craft was in transit; the average orbit, or orbital positions, would be barely an initial step in the calculation that you actually need. Perhaps there is some meaning to the average calculation to that accuracy, and I leave it as an exercise to find such a use. Just because the author doesn't know of any doesn't mean there can't be any.

Problem 3) A fortnight is exactly two weeks long. Find the number of seconds in a fortnight.

Here an estimate is not worth the effort, but the unit conversion is worth looking at. We have 14 days in a fortnight (fn), or a conversion factor of 1.4×10^1 d fn^{-1}. We need seconds, so we have another conversion factor, 8.6×10^4 s d^{-1}. Now we just need to multiply the conversion factor to eliminate the unit we don't need, d: 1.4×10^1 d fn$^{-1} \times 8.6 \times 10^4$ s d$^{-1} = 1.204 \times 10^6$ s fn^{-1}. The conversion factors looked like two SF, but actually they are exact, so we are entitled to the four SF answer.

The d and d^{-1} of course cancel, leaving s fn^{-1} as the units of the answer, as required.

We could continue with these problems to illustrate the principles involved, but we should move on. The following is an exercise, using the results of the previous problem: find the speed of light in furlongs fn^{-1} given that the speed of light is *defined* as 2.997925×10^5 km s^{-1} and there are exactly 8 furlongs mi^{-1}, and 1 km = 1000 m (exactly), and there are 1608 m mi^{-1} (4 SF — this conversion is not exact). Hint: to make sure that you get about the right answer to start, round the speed of light to 3×10^5 km s^{-1}. How do you make the final correction to the defined speed of light (assuming you have a calculator)? Second hint: the answer is of the order of 10^{12} furlongs fn^{-1}.

Estimate: A furlong is 1/8 mi, so 1 furlong = 1/8 mi $\times$ 1600 m mi^{-1} = 200 m. Light speed = 3×10^5 km s^{-1} = 3×10^8 m s^{-1}. One fn = 14 days, and 1 day = $24 \times 3600 \approx 10^5$ s, so to 1 SF, we have 10^6 s, so light speed $\approx 3 \times 10^8$ m s$^{-1} \times 10^6$ s $\times$ 1/200 furlong m$^{-1} \approx 10^{12}$ furlong fn^{-1}.

Answer This has been mainly an exercise in unit conversion. Now it is merely necessary to get out a calculator, and use 1 furlong = 201 m, 1 fn = 14 days (the worst approximation in the estimate was taking this as 10 days), and use the exact speed of light. By taking the correct number of days per fn, we multiply by 1.4; the other changes are small, so we should wind up with 1.4×10^{12} furlongs fn^{-1}, but this is only 2 SF, and, as the only limit on SF is meters per mile, which is good to 4 SF, the answer should be carried out to 4 SF. You are left to finish plugging in the numbers.

COMMENT The estimate in this case was worked out essentially as the complete problem, but to only 1 SF, which did not require a calculator. Since there was only one term that underwent severe rounding off, 14 days per fortnight was rounded to 10, the correction was simple. The only thing we had to notice was that the speed of light, the number of days per fortnight, seconds per day, were all exact (remember the speed of light was *defined* in terms of the given number of meters per second) Only the meters per mile is limited to four SF, and thus limits the number of SF.

3

Part Two: The Standard International (SI) System of Units

In the problems in the first part of this chapter, we used units that are just about never used in science, just to illustrate the way to manipulate units. There is, however, a standard international set of units, and we need to become acquainted with these; we will be using these units from now on.

There are seven fundamental units from which all others may be derived. The seven:

i) Distance in *meters* (m)
ii) Mass in *kilograms* (kg)
iii) Time in *seconds* (s)
iv) Temperature in *Kelvins* (K)
v) Quantity in *moles* (mol)
vi) Electric current in *Amperes* (A)
vii) Light (luminous) intensity in *candelas* (cd)

The first three sometimes cause the SI system to be referred to as *mks*, although there are four additional fundamental units. In chemistry, moles are used constantly, since amount of material is probably one of the most often required quantities in chemistry. Temperature turns out to be the determining factor in equilibrium, and rates, of chemical reactions. Electrical quantities occur in calculating energies of interaction, and in electrochemistry. Probably the least used in chemistry is light intensity, although in photochemistry it is a central quantity.

Obviously, there are quantities that one needs that are not in this list of seven. These are all derived by combining the appropriate quantities from the seven fundamental units. The most obvious example is energy, but let's start with a simpler quantity, velocity = distance/time (*i.e.*, as far as units are concerned, velocity (v) = m s^{-1}). Next, remember that there are multiple forms of energy, one of which is kinetic energy, the energy of motion. For a single particle, kinetic energy = ½ mass × (velocity)^2 which we write as ½ mv^2. The ½ has no units and does not concern us in this context. However, energy is energy, so all forms of energy have the same units, which means that if we get kinetic energy, we have the units for all forms of energy; different forms of energy can be added, so they must all have the same units, whether they are kinetic energy, chemical energy, electrical energy, or any other form of energy.

We have therefore, for all forms of energy, in terms of the fundamental units

$$E_{kin} = kg \times (m\ s^{-1})^2 = kg\ m^2\ s^{-2} = E\ (any\ form)$$

When we come to the chapter on gases (chapter five) we will see that we need several quantities: Pressure (P), volume (V), temperature (T), and number of moles (n). Temperature and moles are fundamental units; we will use n as the abbreviation for moles, Volume is measured by multiplying three distances, so it must have units of m × m × m = m^3. Area is two dimensional, m × m = m^2. Pressure (P) is a little more complicated — it concerns the *force* exerted on a unit area of the gas container. Force, therefore, must also be a derived unit. One thing we know about force is that force × distance = energy. From this:

$$force = energy/distance = kg\ m^2\ s^{-2} / m = kg\ m\ s^{-2}.$$

Then force/area = P = kg m s^{-2}/m^2 = kg m^{-1} s^{-2}. Suppose we multiply this by volume, m^3, to get kg m^{-1} s^{-2} × m^3 = kg m^2 s^{-2} – in other words, energy. We have just shown that PV is a form of energy. This should not be a big surprise, if we think physically; the pressure was the force on an area (*i.e.*, force/area), and the volume is area × distance. The area not only cancels as a unit, but intuitively we can imagine pushing on the unit area of the container wall, so we balance the force; then the distance is the distance we move this area by this push. In other words, it is force × distance, which we knew to begin with was energy. There are other examples of forms of energy that it is instructive to consider, and they all, in the end, can be brought back to the same units. For example, if we go to Newton's Second Law, F = ma, where a is acceleration, we will get the same units for force, and then multiplying by distance will give energy. a is the rate of change of velocity, in other words m s^{-1} × s^{-1} = m s^{-2}, so mass × acceleration = kg × m × s^{-2}, as always for force. Multiply by distance to get kg × m^2 × s^{-2}, exactly the units of energy.

Before we go further, we should note that scientists have a habit of naming things after scientists, so units are commonly referred to by names. Two of the fundamental units are already the names of scientists: Ampere for electric current, Kelvin for temperature. However, the units of energy and force also have names: 1 kg m^2 s^{-2} = 1 *Joule*, (J) and, for force, 1 kg m s^{-2} = 1 *Newton* (N). It turns out that 1 J is a pretty small unit for most of what we need, so we typically talk in terms of kiloJoules (kJ--thousands of Joules). Pressure is given in *Pascals* (Pa). Normal atmospheric pressure is roughly 100,000 Pa (define the unit *atmospheres,* abbreviated atm =101,325 Pa exactly; 1 bar = 100,000 Pa, by definition of bar; many problems are stated in atmospheres, which is a convenient unit). Temperature in *Kelvins* (K) is always positive. There is such a thing as absolute zero, in which the system has its minimum possible energy, and there is no temperature that is lower; temperature is a measure of the energy the system has, but it does not have the units of energy; it is an independent unit. The size of a degree K is the same as the size of a Celsius unit of temperature, but the zero of the Kelvin scale is $-273.15\,^{\circ}\mathrm{C}$ (for Celsius, we write the unit as $^{\circ}\mathrm{C}$, not just C). Many tables of properties of substances are given at $298.15\,\mathrm{K} = 25.00\,^{\circ}\mathrm{C}$. While the temperature is a measure of the thermal energy of a system, it must be multiplied by a constant with the units of (energy/temperature) to convert to energy. The constant per atom or molecule is called Boltzmann's constant, written k_B, so the thermal energy per atom or molecule is $k_B T$. For a mole of atoms or molecules, in place of k_B, we use the symbol R, which is 6.0223×10^{23} times larger, and RT is the thermal energy for a mole of the substance. We will see that one mole of a substance contains 6.0223×10^{23} particles (atoms or molecules) of that substance; this number too has a name, Avogadro's number. (Amadeo Avogadro showed, early in the 19th century, that such a number must exist; his work was understood by the chemistry community after it was explained by his countryman, Stanislao Cannizaro, at a conference half a century later, 1859–60, and the value was finally determined another half century later, in part by Einstein during his "*annum mirabilis*", 1905).

Intensive and extensive properties: Some properties are proportional to the size of the system, like number of moles, or simply volume. Others, however, are scaled by the size so that as there is more of something, say mass, there is more of something else, say volume, that increases in proportion. The ratio of mass/volume is the *density*, and, as is obvious (quantity cancels), so, is independent of the quantity of the material, but is an *intensive property* of the material itself. Temperature is also an intensive quantity Volume and mass depend on the quantity, and are thus extensive properties.

Dimensionless variables: This is generally a more advanced topic, and not encountered in first year chemistry, so we will just mention it. In some cases, it is useful to redefine the variables in such a way that the dimensions cancel. Suppose we consider an equation that we will come to in Chapter Five, the van der Waals equation for a non-ideal gas (here written for one mole):

$$(P + a/V^2)(V - b) = RT$$

where the a and b terms are small (usually) corrections to the larger terms of the ideal gas equation, which, for one mole, is PV = RT. Obviously the small correction constant **b** has dimensions of volume, and **a** has dimensions of pressure × volume^2; it is not possible to add two quantities with different dimensions, so these dimensions are required to make the first parenthesis consistent with dimensions of pressure, the second with dimensions of volume. Expand the left hand side: PV + **a**/V −**b**P = RT; we throw away the doubly small −**ab**/V^2 final term; with **a** and **b** both small, their product is often too small to worry about (not always). Now all terms have the dimensions of energy. Divide through by RT, and define x = PV/RT, which is dimensionless. Then the equation becomes x + **a**/RTV − **b**P/RT = 1. In this case, it is not particularly clear that this helps, but possibly, with both P and V in this form, one more step is appropriate; use P = (RT/V) x, leaving

$$x = 1 - a/VRT - b/V = 1 - (1/V)(a/RT - b)$$

As an exercise, check that each term is dimensionless; if it were not, this would be a nonsensical expression. If the magnitude of **a** and of **b** is small enough, one could carry out successive approximations starting with x = 1 (the ideal case), and get a good result more quickly than by solving the original equation. Alternatively, one could plot x vs 1/V. and perhaps put V in terms of the volume of a mole of ideal gas (see Chapter Five; here just take it to mean x = 1), at *standard temperature and pressure* (V_o) (STP: 273 K, 1.01325 × 10^5 Pa, giving V_o = 0.0224 m^3: for V, the value of R, a universal constant (0.0821 L-atm mol^{-1}K^{-1}), is used to get this). Introduce the variable *V* defined as V/V_o; this changes the units of V to dimensionless, but does not affect the rest of the equation. Such a plot would give a sense of the importance of the correction term, which would be relatively large if $1/V$ were large, but small if $1/V$ were small. This is not a topic to which we shall return, but for completeness, we mention it once here.

The possible problems for this section are essntially the same as those at the end of the previous section, and the same principles apply.

Problems:

1) A student finds an exam question that requires the Ideal Gas Law, PV = RT for 1 mole) but her mind has gone blank, and she can't remember it. However, at the bottom of the page there are some constants needed for the numerical problems, among them the gas constant R, shown as .0821 L-atm $mole^{-1}K^{-1}$. The student was saved: using this set of units, an equation with consistent units can be created. For a number of moles other than 1, cancel L – atm by dividing by PV, and multiplying $mole^{-1}K^{-1}$ by nT, making

   R (nT/PV) =1, or, multiplying through, one gets

$$PV = nRT$$

   What remains is simply to plug in values. How do you know the answer cannot have more than three SF?

2) The density of water at 25°C is 0.997 g cm^{-3} = 0.997 g mL^{-1}. What is the density of water in lbs ft^{-3}? The foot is exactly 12 inches, and the inch is now defined as exactly 2.54 cm. One lb = 453.59237 g.

   Estimate: We have two conversions, to begin with: g to lbs, and cm^{-3} to ft^{-3}. We see at once that there are a lot more g than lbs. In round numbers, let the 453.6 g be about 500 g, or the 1 g is around 0.002 lb. Also, the conversion of cm^{-3}

can go through two steps: cm $\to$ inches $\to$ foot. 2.54 cm in^{-1} × 12 in ft^{-1} ≈ 30 cm ft^{-1} Therefore, $(30)^3$ cm^{3}ft^{-3} ≈ 2.7 × 10^4 cm^{3}ft^{-3}. So 1 g cm^{-3} × 0.002 lb g^{-1} × 2.7 × 10^4 cm^{3}ft^{-3} ≈54 lb ft^{-3}. A couple of notes: In setting up the final line of the estimate, we canceled the units cm^{-3} × cm^3 and g × g^{-1}. If we had made a mistake, and put the factors in the wrong order, we would have seen this immediately as the units would fail to cancel. Also, we could tell if the number were really unreasonable. Suppose we had the .002 value inverted, getting > 10^7 lbs ft^{-3}. This is so obviously absurd that we should pick it up even without much thinking. We have all picked up something like a cubic foot of water (maybe a bucket's worth), and it did not weigh 5000 tons. Inverting the other factor would produce an equally absurd answer, this time way too small. The point of this exercise is not simply to go through the unit conversion, but to show how the numerical value can easily become absurd, and that we should pay attention to that value as well.

Answer: We are about done — just put in 0.997 in place of 1 g cm^{-3}, 30.48 cm ft^{-1}, and 453.6 g lb^{-1}. The 2.54 cm inch^{-1} is exact, as is the 12 inch ft^{-1}, so the 30.48 cm ft is exact. To get the final answer it is worth carrying a fourth SF, so that error doesn't pile up. We only have three SF, as the 0.997 has only three figures, so we must round off at the end. We already have the correct form, so it is just a matter at this point of plugging the numbers into a calculator. The estimate is worth doing to make sure that we won't get a crazy answer, but also to make sure that the units are correct. This problem mainly illustrates the use of cancelling units. The final numerical answer is 62.2 lbs ft^{-3}. The main difference from the estimate comes from the approximately 10%

round off of 453 to 500. There is also about 4% from the round-off of 30.48 to 30 (remember this is cubed, so it is about triple the one dimensional round off). As long as we are paying attention, we might have estimated in advance that there should be about this much error from the round off, and in this direction. This elaborate discussion would not normally be necessary, but it's worth going through here to see what it is important to think about. In future examples, we may not bother with examining the estimate so closely.

3) Often, environmental problems call for 1 SF answers, with very large volumes combined with very small concentrations. Here is a problem in which we can make use of the conversion factors from the previous problem. Problem: On a certain day in 1987, the concentration of carbon monoxide (CO) in the air over New York City was measured at 2.0×10^{-6} g L^{-1}. If the total volume of air over the city is 1.5×10^{14} ft^3, how many English tons of CO were there over the city that day? (An English ton weighs 2000 lbs, about 10% less than a metric ton).

Estimate: We have to make two unit conversions, L $\rightarrow$ ft^3 for the volume, and g$\rightarrow$tons for the mass. As in the previous problem, we convert the L = 1000 cm^3 to ft^3 using the 30.48 cm = 1 foot,

i) 2.7×10^3 cm^3 $\approx$ 1 ft^3, and as 10^3 cm^3 = 1 L, 2.7 L $\approx$ 1ft^3

ii) To go from g to English tons, 1g $\approx$0.002 lb $\times$ 1/2000 lb ton^{-1} = 1×10^{-6} tons.

We can now set the cancellation of units: Therefore we have

$$1.5 \times 10^{14} \text{ ft}^3 \times 2.7 \text{ L ft}^{-3} \times 2 \times 10^{-6} \text{ g L}^{-1} \times 1 \times 10^{-6} \text{ ton g}^{-1} \approx 0.8 \text{ tons.}$$

Again, units cancel appropriately: ft^3, L, and g all cancel, so we have set up the problem correctly. It looks like it doesn't take much to pollute a big city. 0.8 tons, or 1600 lbs, would do this much. This concentration of CO is a little less than it would take to give everyone in the city a serious headache, at least. Next question: is it worth going beyond this estimate to get a more accurate result? Here is where our common sense comes in. Is it likely that the concentration is constant over such a huge volume? Surely not. The concentration is likely to vary from place to place, and the value given can only be an average. This is the kind of problem where the estimate is about the best we can do. (and by the way, does 10^{14} ft^3 for the volume of air sound reasonable? Consider the size of the city.).

Summary We can use any units we want, and the principles for manipulating them are the same. However, there is a standard set of units that we have met in the second part of this chapter, and these are the ones that are used in science, so we have to get used to using them. There are times when we will use a non-standard unit for convenience, but these units are traceable to SI units. For example, consider atmosphere as a unit of pressure. It is defined in terms of SI units as 101,325 Pascals. Also, atomic mass units are used for the masses of atoms and molecules, which are tiny compared to kg. However, for the most part we will deal with either the standard units, or combinations of them.

4

Atomic Mass and Molecular Mass

Atoms and molecules are composed of particles that have a definite mass. These used to be referred to as atomic weight and molecular weight, but this is no longer considered the proper notation. Weight is proportional to mass, the strength of the gravitational field being the proportionality constant — except that it isn't really constant, because it depends on where you are. If you are circling the earth in a satellite, there is zero weight. If you are on a planet with a stronger gravitational field, the weight will be greater. The mass is unchanged, however, so it is better to refer to mass, and that is the proper scientific way to consider it.

Atoms are made up of protons, neutrons, and electrons. The protons and neutrons comprise almost all the mass of the atom; the electron has almost no mass (one electron is about 1/1800 the mass of a proton or neutron). We can define an atomic mass unit (amu) that sets the mass of a proton or a neutron to approximately one amu each, add up the number of protons and neutrons, and call this the atomic or molecular mass; however, there are complications, so this is only approximately the atomic mass unit definition (we will get to the actual definition below). Obviously amu is not close to kg.

One complication is that there is a *strong force* that holds the atom together. Like charges repel, the protons and neutrons are in a very compact *nucleus* of the atom (the electrons are distributed in a much larger space, not simple orbits like planets around the sun, but over a space that may take complex shapes). The protons therefore have to be held together, and this is the function of the strong force; it carries an energy, so it adds a little to the mass.

The protons have a positive charge, the electrons have an exactly balancing negative charge, and the neutrons have no charge. The number of protons, called the *atomic number*, defines which atom it is: hydrogen has 1, helium has 2, lithium has 3, uranium has 92, and so on. The number of neutrons is not fixed for a given type of atom. Typically for light atoms the most common number, if the number of protons is even, is the same as the number of protons, but other *isotopes* exist (isotopes of a single element). However, the number of neutrons may vary, and each isotope of a given element will have a different number of neutrons: for example, ^{11}C — carbon 11 — has 5 neutrons, ^{12}C, the most common isotope, has 6 neutrons, ^{13}C has 7, ^{14}C has 8. The most stable configuration is ^{12}C and this makes up about 99% of the carbon found on earth, with ^{13}C being the last 1%.) If a light atom has an odd number of protons, there may be one more neutron than proton, in most cases, although odd-odd nuclei exist. Heavier atoms have more neutrons than protons. The electron distribution is spread out over a volume with a radius of the order of 10^4 times the radius of the nucleus. The strong force translates to an energy, and remembering that $E = mc^2$, there is a mass dependent on the forces in the nucleus, so the masses of the atoms are not exactly simple multiples of the mass of protons and neutrons, although typically they are close. As a result, we have to pick a particular nucleus, or rather an atom, to have a defined mass on the atomic mass scale; the one chosen is the most common isotope of carbon, with six protons and six neutrons, defined to be exactly 12; ***this is the definition of the amu scale***. All other atoms have a mass relative to this isotope. As we said earlier, the mass of the atom is only approximately the mass of the number of neutrons plus protons of which it is composed.

Molecular mass: The molecular mass is the sum of the atomic masses of which the molecule is composed. For example, the mass of carbon is 12.01 (the carbon isotope of mass 13 is 1% of the carbon on earth and the atomic mass that we use is based on the average distribution of isotopes on earth; ^{13}C is responsible for the atomic mass of carbon not being exactly 12). The atomic mass of oxygen is 16.00, so the mass of carbon monoxide, CO, is $12.01 + 16.00 = 28.01$. The mass of carbon dioxide, CO_2, is $12.01 + 2 \times 16.00 = 44.01$. To get the mass of a molecule on the amu scale, we simply add the masses of its atoms. In some parts of chemistry, especially biochemistry, only relatively light atoms, especially C, N. O and many H atoms, are involved. Phosphorus (P) and sulfur (S) are also important in biology. However, much of chemistry is not biological; most metals are heavier than these. Iron has mass 55.85, and a major component of iron ore, Fe_2O_3, has a molecular mass of $2 \times 55.85 + 3 \times 16.00 = 159.70$. Uranium has an atomic mass of 238.03, so U_3O_8 has a molecular mass of 842.09 (work this out as an exercise). Note that the atomic masses are often close, but not all that close, to integers, partly for the reasons we have noted earlier; the major reason for their being non-integer comes from the mixture of isotopes. If you have a heavy atom, its contribution forms a quick lower bound when the other atoms are light. In U_3O_8 just taking 3×238 sets a lower bound, for example. However, molecular mass is not very difficult to calculate, so just getting the mass by adding the atomic masses is simple enough that an estimate is rarely worth the effort — as long as you check that the result you arrive at is not ridiculous (e.g., if you punched the wrong button on your calculator).

Isotopes: We defined an isotope above to mean the same kind of atom — same number of protons and electrons, which we will see later means essentially the same chemistry — but a different number of neutrons. What is changed is the mass of the atom, and it need not be changed by exactly an integer, in part because of the strong force holding the protons and neutrons together, which changes a little when a neutron is added or subtracted. The atomic masses are defined in terms of their mass relative that of the isotope of carbon with six protons and six neutrons, which, as we just noted, is defined as exactly 12. The mass of a normal sample of carbon, however, will have a slightly higher mass, because it contains a little of the isotope with six carbons and seven neutrons, which would have a mass of about 13 (not exactly — only carbon 12 is defined as exact). We just saw the atomic mass of carbon is 13.01. One of the problems below starts with this as an estimate; the value on the big periodic table in the front of your lecture hall (if it is like most chemistry lecture halls) is given as 12.011, with one more SF. We will talk about this with a worked out problem later in this chapter. Hydrogen's most common isotope has a nucleus that is just one proton, but also has a rare, but stable, isotope, deuterium, with a nucleus with one proton plus one neutron, so a sample of hydrogen will have a mass slightly greater than one, 1.008. Stable means that it does not decay to produce a different element — we discuss *radioactivity* in the last chapter. Other atoms also are not close to integers. Sometimes there are fairly equal numbers of two different isotopes; with chlorine, one has about 76% isotope 35, 24% isotope 37, and the overall mass is thus about 35.45. The isotope 36 has odd numbers of both protons (17) and neutrons (19), and is radioactive. Odd – odd nuclei tend to be unstable.

The masses we use in standard chemical calculations are those from a sample of the atoms. What kind of sample? One from planet earth, as that is all we can get. The abundances will not be very different elsewhere in the universe, as the energy of a nucleus is the same everywhere, and the conditions under which the nuclei form are probably not so tremendously different — but they may be a little different, so we must remember that our atomic masses depend on an earth-centric sampling. Even on earth, if you carry out the averaging to a couple of extra SF, not all samples are quite identical. Having noted this, we will never return to it, and just proceed with the abundances we have.

There are a number of questions that we can begin with. For one thing, we would like to convert amu to kg, or grams, which is the more common unit in practice in this case (kg remains the SI unit, of course). To begin with, we know that there are a **lot** of atoms in a gram. If we consider the carbon 12 isotope, then the ratio of 12 g/12 amu must give the standard number of atoms in one **mole** of atoms; this is Avogadro's number, mentioned earlier in part 2 of the chapter on units. In both cases it is a matter of getting the number of atoms in a mole. Sometimes in the first exam in first year chemistry, you are asked to find the mass, in grams, of a hydrogen atom (atomic mass = 1, approximately). Since it takes Avogadro's number of these atoms to make 1 g, each atom must weigh $1/6 \times 10^{23}$ g, about 1.6×10^{-24} g. (Multiplying the amu by Avogadro's number instead of dividing produces the error referred to at the beginning of the book, giving something like the mass of the moon instead of the mass of the hydrogen atom.)

SUMMARY: Atoms, and the molecules they form, have mass. This mass is usually expressed in atomic mass units, a scale on which naturally occurring atoms range in mass from 1 to about 238, and artificial heavy atoms can be somewhat more. Molecules on this scale have mass equal to the sums of the masses of which they are comprised. If there are not too many atoms, and not very heavy ones (e.g., carbon, nitrogen, oxygen, or even sulfur), we expect a molecular mass in the range of maybe 50 or so up to maybe 1000. Metals are generally heavier (although lithium is very light). If you multiply the number of atoms (except for hydrogen) by 100, you will come up with a number 100 times the number of metal atoms, and you will have something like the correct mass. For an organic compound with only carbon, nitrogen, and *oxygen, plus hydrogens, it should be around 15 times the number of non-hydrogen atoms, after you add in the hydrogens. If the question asks for the mass in grams, **divide** the amu value by Avogadro's number. Finally, watch out for complications involving isotopes (these make good exam questions, even if they are not usually all that important in real life). Be careful also of significant figures; errors can build up if there are large numbers of atoms to be added together. Only the isotope carbon 12 has an exact integer weight.*

Problems:

1) Argon has three naturally occurring isotopes, with abundances as follows: 0.337% ^{36}Ar, mass 35.968, 0.063 % ^{38}Ar, mass 37.963, and 99.60% ^{40}Ar mass 39.962. Find the atomic mass of argon.

   > *Estimate*: Looking at the percentages, it is obvious that the mass will be almost that of ^{40}Ar, in other words a little less than 40. Looking at the masses, it is clear that the mass of that isotope is an upper bound, as the others have less mass, although they don't matter until the third SF, at least. We should expect a mass around 39.9.
   >
   > *Answer*: atomic mass = .00337 × 35.968 + .00063 × 37.963 + .9960 × 39.962 = 39.947

   Since the abundance of the major isotope is given to only four figures, we should round this to 39.95. The masses of the minor isotopes could have been rounded to two figures for ^{36}Ar. If we had totally ignored ^{38}Ar, would it have mattered? (check this yourself).

2) The atomic mass of carbon is 12.011. Naturally occurring carbon consists of two isotopes. The isotopic mass of ^{13}C is 13.003355. Find the relative abundance of the two isotopes.

Estimate: We really don't have the information we need to do this problem, so let's make the assumption that the only other isotope is ^{12}C with a mass of exactly 12. The atomic mass is about 12.01, about 1% of the way from 12 to 13, so we expect about 99% ^{12}C, and 1% ^{13}C.

Answer: Call the abundance of ^{12}C z, and that of ^{13}C 1-z. Then $(1-z) \times 13.003355 + z \times 12 = 12.011$, and solve for z. The abundances are then ^{12}C = 98.9%, ^{13}C = 1.1%.

NOTE: It is only for carbon 12 (^{12}C) that it is legitimate to make the assumption that the isotopic mass is exactly an integer, 12. Actually, given the information in the question, even the assumption that the lighter isotope is one less is not justified, and there are atoms with odd atomic numbers, such as Cl, where such an assumption is false. We just happen to know that for carbon, the main isotope is ^{12}C. Also, why do we have so few SF in the answer? There were more in the masses that were given. Here, the small difference in two large numbers limits us. In solving the equation, we had to take differences in mass that required an intermediate result with fewer SF.

3) We mentioned chlorine isotopes in discussing the previous problem. Let's do essentially the same problem, but for Cl this time. Given: mass of ^{37}Cl = 36.96590, of ^{35}Cl = 34.96885, and the atomic mass of natural Cl = 35.453. Find the isotopic abundances.

Estimate: Looking at the masses, we see essentially 37 and 35, and the average around 35.5, so right away we see that there must be more 35 than 37, and roughly by 3:1 since the 35.5 is about ¼ the way from 35 to 37. There must be approximately 75% isotope 35, 25% isotope 37.

Answer: We know how to set this up: let the abundance of 35 = z, of 37 = 1 − z, and do exactly what we did in the previous problem. Solving gives abundance of 35 = 75.76%, of 37 = 24.23%. Again we have to be careful with SF, but the masses are given to seven places, while only the final atomic mass is given to only five places, so we do 1 SF better than with the carbon question above.

COMMENT: The abundance of each isotope might vary slightly from one sample to another so we have to careful with our SF for this reason as well.

4)   Find the molecular mass of $YBa_2Cu_3O_7$

Estimate: We have six metal atoms and seven oxygen atoms. If we take the average metal atom as 100, and 7 oxygens as another 100, we get $6 \times 100 + 100 = 700$. This is a pretty rough way to do an average, but it is surprisingly useful. Metal atoms in general (neglecting lithium) range from around 25 to around 200, but we will find that we get in the right range with this very rough estimate.

Answer: The atomic masses are Y = 88.91, Ba = 137.33, Cu = 64.46. O = 16.00

So molecular mass = $1 \times 88.91 + 2 \times 137.33 + 3 \times 64.46 + 7 \times 16.00 = 668.9$

We could in principle carry this one more figure, but uncertainty in the last figure of the atomic masses, plus round-off error, would really make the last figure uncertain. There are 6 metal atoms, plus 7 oxygens, so there is enough uncertainty to justify rounding off after the first decimal place. This happens to be a particularly interesting compound; slightly oxygen deficient forms are among the first "high temperature" superconductors. While estimates for this type of problem are rarely done, and for that matter, rarely useful on their own, they do have a use as part of a larger calculation.

5

Gases

Gases are a state of matter in which the molecules are far apart, and behave almost independently of each other. We can define an "Ideal Gas" in which the molecules are effectively points with mass; they can collide, so they can exchange energy and thus reach equilibrium with a uniform temperature (T) and pressure (P) throughout. They are enclosed in some sort of container, so that their volume (V) is defined, and the number of molecules (the number is huge so usually we consider this as the number of **moles** (n), which can be of order 1, although several orders of magnitude difference are possible if the system is much larger or smaller). However, there is no way that the number of moles even in a large system is going to be anything like as big as the number of molecules in even a small system). For an *ideal gas*, if we know three of the variables (P, V, n, T) we can find the fourth — the ideal gas is defined by following the Ideal Gas Law, as defined below. We used the dimensions of these variables as examples in the chapter on units.

If the molecules become very crowded (the volume, for a given number of moles, gets small), they hit the wall more often, and the pressure goes up. The pressure is produced by the molecules hitting the wall; if there are more molecules, or they hit harder, the pressure goes up. How would they hit harder? The energy of ideal gas molecules is just the energy of motion (kinetic energy), and the measure of average energy is temperature. Higher temperature means the molecules move faster, so they hit the walls harder, so the pressure goes up. In other words, we should expect the pressure to increase if the number of moles goes up, the temperature goes up, and the volume goes down. Sure enough, for an ideal gas, we can relate these quantities as

$$P = nRT/V$$

just as our qualitative discussion suggested, with R being the constant of proportionality that makes the units come out right, and at the same time makes the numerical values fit together in units that are consistent with those we use for everything else. In principle, in SI units we should have P in Pascals ($= N\ m^{-2} = kg\ m^{-1}s^{-2}$), n in moles, T in Kelvins, and V in m^3. Then R must have units $Pa \times m^3 \times mol^{-1} \times K^{-1}$. These units turn out to be inconvenient; most of the time m^3 is too big, so we use liters $(L) = 0.001\ m^3$, and Pascals too small, so we use bars; 1 bar is 100,000 Pa. Sometimes (often) it is more convenient to use *atmospheres* (atm); 1 atmosphere $= 101,325$ Pa, which is about the standard atmospheric pressure at sea level. Pressure has historically been measured with a mercury barometer; atmospheric pressure can support a column of Hg, and the standard atmospheric pressure corresponds to a column of 760 mm Hg. If we use L for volume, and atmospheres for pressure, $R = 0.0821$ L-atm $mole^{-1}\ K^{-1}$. As we saw in the introductory chapter on units, pressure x volume is energy, so R must have units of energy $mol^{-1}K^{-1}$; in these units, $R = 8.31$ J $mol^{-1}\ K^{-1}$. These units are not convenient in this chapter, but later we will use them. It is worth remembering that at standard temperature and pressure, STP (T = 273 K; P = 1 atm) the volume of one mole of ideal gas is 22.4 L. This works out to about 25 L (actually, 24.6 L to three SF, but for an estimate, 25 L is useful) around 300 K (many standard tables of values use 298 K = 25°C). This helps keep the answers to ideal gas problems anchored.

For an example: Suppose we have a sample at 500 K, with 0.173 moles. If the temperature is 500 K, less than twice 300 K, the molar volume at 1 atm would be less than 50 L. The number of moles is 0.173, which is less than 0.2 moles, so the STP volume is less than 5 L, and at 500 K somewhere around 8 L. We won't go through the complete solution. Simply by remembering the STP volume, or just 25 L at one atmosphere pressure, 300 K, one can check very quickly whether the answer to the particular problem is reasonable or ridiculous. We will have a few examples at the end of chapter.

What happens if the gas is far enough from ideal to require correction? There are two ways this can happen. There are significant forces between the molecules, and the space occupied by the molecules is not negligible. First of all, this happens if the molecules are fairly densely packed, for a gas — still not tightly packed as in a condensed phase, but perhaps the molecules take up 3% of the total volume, enough so that a calculation to three figures would have to account for this. The other thing that happens is that the molecules are close enough, often enough, to attract each other. This means that both the pressure is decreased because the molecules are pulling each other in, away from the walls, and the free volume is decreased by the space occupied by the molecules so the distance between them is reduced so much that their force fields interact. Now we modify the ideal gas law by accounting for these two effects, by including two new small constants. These are not universal constants, but depend on the gas. We wrote the ideal gas law for one mole, but let us include the number of moles when we write the *van der Waals equation.* Actually, we met this equation in the units chapter, but there it was written for one mole; here, it is written for any number of moles, where n = moles. This is not the only equation for non-ideal gases, but it is probably the most popular — there are other algebraic forms that express the same physical facts.

$$(P + a\,(n/V)^2)(V - nb) = nRT$$

Obviously if $a = b = 0$, we are back to the ideal gas equation, and with $a = 0$, $b > 0$, the effective volume is reduced by nb. When it comes to estimating values we should realize that under any reasonable conditions a and b are small, and will change the final answer somewhat, but the general rules for the range of plausible answers are pretty much the same as for the ideal gas. This said, the van der Waals equation, if we go on to higher densities, keeping the $a \times b$ term (making it a cubic equation) can describe much more complex behavior of fluids that are much denser than what we usually think of as a gas, continuing on to liquid-like behavior; the cubic equation shows how a phase change can be expressed. However, describing a phase change is too advanced a topic for an introductory course. When it comes to actually solving the equation, in the units chapter we suggested expanding it in terms of the small constants, and neglecting the second order term with $a \times b$. Other forms for non-ideal gas equations also amount to expansions like this; one is called the Virial expansion. We will reserve further comments for the problems themselves.

If you have a mixture of ideal gases, we expect that the pressure of each gas is determined by the number of moles of that gas at the given temperature and volume. The total pressure is just the sum of the pressures of the individual gases; this is Dalton's Law of Partial Pressures. This follows from what we have said already; the pressure is due to collisions of molecules with the wall, and the molecules do not interact, so the collisions of each type of molecule simply add up. The Law is normally expressed in terms of *partial pressures*, with the partial pressures proportional to the mole fractions of the individual gases; the mole fraction of a gas G is defined as $x_G = n_G/n_T$ where n_G = number of moles of gas G, and n_T is the total number of moles of all gases present. The partial pressure of G is then $p_G = x_G\,P$, where P is the total pressure of the mixture. Since the sum of the mole fractions equals 1, this means that sum of the partial pressures equals the total pressure.

This seems almost obvious, but it certainly was of historical importance, as it led Dalton to infer the Law of Multiple Proportions: if two elements formed more than one compound, the amount of each could be attributed to the number of atoms of each in the compound — in other words, atoms must exist. Dalton, of course, is known for having established that there must be atoms early in the 19th century, and before Avogadro's work (it took about another century to convince the last of the doubters about the existence of real physical atoms, although the chemistry community as a whole got it right in the 1859–1860 conference we mentioned above, when Avogadro's work was explained by Stanislao Cannizzaro to a conference called to sort out the confusion. Mendele'ev was at the conference, and used what he learned, with other information, to put together the Periodic Table.).

Everything we have discussed up to this point concerns the static behavior of gases. Things can change with time. Later, we will consider rates of reactions, but here we can introduce one form of time dependence, in which there is a small leak in the apparatus, so the gas pressure drops with time. The gas *effuses* at a rate proportional to the inverse square root of its molar mass.

$$\text{Rate } \alpha \text{ } M^{-1/2}$$

where α is read as "is proportional to". This is Graham's Law. If we have two gases leaking, they separate, since they effuse at different rates:

$$r_2/r_1 = M_2^{-1/2}/M_1^{-1/2} = (M_1/M_2)^{1/2}$$

r_2/r_1 is the ratio of rates of effusion. If $M_1 > M_2$ then $r_2 > r_1$; because of the inverse ratio of rate to mass, the lighter gas effuses faster.

When it comes to the mixing of gases, or the rate at which molecules of a gas move through a mixture, called diffusion, again the lighter gas moves faster, according to the inverse square root of mass. Rates of diffusion are also related to the rates at which the molecules collide, and thus to reaction rates. However, this is a topic for a later chapter.

SUMMARY: In doing problems on gases we are able to anchor our answers by the known value of the molar volume at STP. Molecules of gases hardly interact, and the behavior of gases is often approximated as an "ideal gas" in which there is no interaction except for the collisions that establish equilibrium among the molecules. In doing calculations, it was convenient to switch units; working with SI units directly would have led to many extra orders of magnitude that would be difficult to keep track of. As to the general magnitude of the answers, knowing the molar volume helps set limits on what we can expect, even if we look at the problem of the pressure of hydrogen in interplanetary space. In that case we know we should have a very low pressure, so if we somehow find a pressure fairly close to the pressure at the surface of the earth, we know the answer must be way off. On the other hand, if the quantities — pressure, volume, temperature — given in the problem are fairly close to earth normal, and we get an answer that is orders of magnitude different, we should similarly know that something is very wrong. In this chapter, as in others, a knowledge of a couple of orders of magnitude, or better, specific values like the molar volume at STP, helps us keep our answers in bounds. A physical sense of what it means to one variable when another variable is altered is necessary. Increasing the pressure makes the volume smaller (this seems essentially intuitive), while raising the number of moles or the temperature should increase the volume, if the other variables do not change. Once we keep these kinds of considerations in mind, we will not multiply when we should divide, or make similar blunders. We should also remember that some quantities are much smaller than others. Do we need to use the van der Waals equation, or is the ideal gas a good enough approximation? In answering this question, we need to consider how many significant *figures are required. This determines whether the volume is larger by enough that we can neglect, for example, the nb term in the van der Waals equation, to that number of significant figures, and similarly for the [a (n/V)^2] term. The estimate can usually be made from the ideal gas law, even if a final answer requires the complete non-ideal equation. A computer may do the work to complete mathematical accuracy, but the supposed added accuracy may not be real, because some input quantity is not known to enough accuracy. As usual, if we consider what our answers mean physically, we will avoid errors. Once we have this much, we can consider the behavior of mixtures of gases, and a little about the rates at which gases can mix and leak through a pinhole.*

Problems:

1) A sample of Cl_2 is held in a 125 mL vessel at a pressure of 55 mm Hg. It is allowed to expand to a volume of 550 mL, at constant temperature. Find the final pressure.

> *Estimate*: First of all, we assume that the gas is ideal, since we are given no additional information; also, "a sample of" means the amount of gas does not change, that is, the number of moles is constant. We cannot check non-ideal behavior, given the lack of relevant information. In any case, for the estimate we would use the ideal gas law. This said, with the temperature and number of moles constant, we only have to use the constancy of the product of P × V under these conditions. V increases by more than a factor of 4, so the pressure must drop by more than a factor of 4. Without doing any arithmetic, we can quickly see that 14 mm Hg pressure is an upper limit for the answer, and probably not that far off
> *Answer*: 125/550 × 55 mm Hg = 12.5 mm Hg

This is a fairly trivial example, but it allows us to see how to proceed.

2) A steel vessel, volume 3.00 L, that can withstand a pressure
    of 10.0 atm holds 3.00 atm of an ideal gas at 300 °C (consider
    this to also be three SF). It is heated at constant volume. At
    what temperature does it burst?

    > *Estimate*: Here, n and V are constant. Pressure is thus
    > proportional to temperature. To begin with, we need to
    > switch from temperature (°C) to temperature (K), so add
    > 273 to the temperature. In round numbers, add 300.
    > Therefore, we have to increase the temperature from
    > about 600 to 10/3 × 600, or about 2000 K. The 10/3 is
    > the ratio of 10 atm (bursting pressure) to the original
    > 3.00 atm. Since we added too much in converting from
    > Celsius to Kelvin temperature, this is an upper limit.
    > *Answer*: We only have to make the initial temperature
    > exact, 300 + 273 = 573 K, so the final temperature is
    > T = 10.0/3.00 × 573 = 1910 K, or 1637°C.

COMMENT: Again, why bother with what seems to be a trivial
problem? For one thing, it gives us an excuse to ask whether this
is a physically realistic problem. The steel will weaken as it is
heated, and ordinary steel melts around 1300°C (pure iron, 1535°C),
well before the temperature that the nominal steel vessel will
burst. Therefore, the problem is not physically reasonable.
Tungsten (m. p. 3410°C) might be realistic. Second, what about
SF? We reported 1637°C, or 4 SF. We really should round off to
three SF, as we don't know the starting temperature to better than
three SF, making the answer 1640°C.

3) Part 1: 7.00 L of H_2 at STP is cooled to its condensation tem-
    perature, 20.7 K. Find the volume at that temperature, if the
    pressure and volume are unchanged.

    > *Estimate and answer.* Again this is a trivial problem, just
    > like the two earlier problems, and we observe very
    > quickly that the final temperature is appreciably less
    > than 1/10 the original temperature, so the volume must

be likewise <1/10 the original volume, hence < 0.70 L. The final volume is 0.531 L.

Part 2: Next, suppose that the pressure is lowered from the STP value (1 atm) to keep the H_2 in the gas phase at 18 K, below the 1 atm condensation temperature (i.e., the boiling point drops as the pressure drops, so, if the boiling point at 1 atm is 20 K, and we want to lower the pressure so that it remains in the vapor phase at 18 K, how much of the gas must be removed or added to make the boiling point 18 K? At 18 K, hydrogen boils at 0.741 atm. *Estimate*: As stated, this is a confusing problem. What is really asking is how to adjust conditions so that the hydrogen remains near the liquid-vapor boundary at 18 K and 0.741 atm. There is a small pressure drop from STP, but a big temperature drop. PV is little changed, so nRT must also remain little changed. R doesn't change, T drops a lot, so n must rise correspondingly. With this understood, we can proceed. This is no longer so trivial a problem. We can use the complete Ideal Gas Law, and observe that the volume remains constant, so we don't need that for the ratio. For an estimate, we see at once that the pressure has dropped to 0.741 atm, roughly the same order of magnitude. The temperature is roughly 1/15 the initial STP temperature, so to keep nearly the same pressure, we will need many more moles. The volume remains 0.531 L. Because the temperature drop is the dominant effect, the estimate is around 15 times as many moles, with the understanding that this is an upper limit, with the pressure requiring some compensation. *Answer*: Number of moles, n = PV/RT

Take the ratio of n(initial)/n(final): Volume and the gas constant R cancel, so n(initial)/n(final) = [P(initial)/T(initial)]/[P(final)/T(final)]

It is more convenient to rearrange

= P(initial)/ P(final) × T(final)/ T(initial)

Now a numerical estimate is easy: T(initial)= 273 K

n(initial)/n(final) = (1/0.741) × 18/273 = .0890

Therefore we need almost 11 times as many moles to have 0.741 atm at 18 K as for 1 atm at 273 K.

COMMENT Suppose we found that n(initial)/n(final) was >1. This would be physically unreasonable, as it would mean we would have to have pumped gas **out**; if we have a big temperature drop, and no other large change, it follows that n must rise accordingly. If the concentration (pressure) gets larger, then the boiling point would rise, not fall, but maintaining pressure near 1 atm at a much lower temperature requires more moles to maintain equilibrium at the new boiling point. This is the kind of check that an estimate is good for. Did we really need to set up the equations? It is faster if we do. If we try to think through this problem, without writing down the equations, for most of us, it would take longer. Nevertheless, the direction of the answer is obvious, as maintaining nearly the pressure requires more moles at the much lower temperature, and checking that the ratio n(initial)/n(final) << 1 really requires checking the arithmetic.

COMMENT: This is a problem in which it is easy to get confused, because normally one looks at the drop in pressure required to lower the boiling point; here, we are concerned with a third temperature, the starting STP value, and are comparing the number of moles at STP to those at the lower temperature. If we compare the number of moles at 20.7 K to the number of moles at 18.0, there would be a ratio of $(P_1/P_2)/(T_1/T_2) = (0.741/1.00)/(18.0/20.7) = 0.644 < 1$.

4) In some parts of interstellar space, there is, on average, 1 H atom/cm^3. Suppose the gas is in equilibrium with the universal thermal background temperature, about 2.7 K. What is the pressure of this gas?

Estimate: This is still an ideal gas question, but with numbers that are chosen to challenge our intuition. Still, we can use pV = nRT. The main thing is that n seems pretty minuscule; in a gas at STP, there are a mole of molecules in 22.4 L, so around 6×10^{23} / (22.4 × 1000) per cm^3 (the factor of 1000 from 1000 cm^3 L^{-1}) or around 3×10^{19} cm^{-3}. This means that we should expect about 19 orders of magnitude lower pressure; however, the temperature is also two orders of magnitude lower than at STP, and this works in the same direction, so we expect around 21 orders of magnitude lower. In other words, we look for pressure around 10^{-21} atm, or bars — in this case it hardly matters.

Answer: Go back to p = nRT/V = (10^3 atoms L^{-1}/6 × 10^{23}) 0.0821 L atm mol^{-1} K^{-1} × 3 K/1 L

= 4×10^{-22} atm. This is in pretty good agreement with the estimate.

COMMENT: How likely is it that the temperature is really all that uniform over galactic distances? Actually, pretty likely, as it is the cosmic microwave background from the Big Bang, uniform to within a small fraction of a degree. Also, the gas is H atoms, not H_2 molecules. Later in this book we will get to equilibrium constants and see how this makes sense.

5) A pressure vessel containing 100 kg of O_2 develops a pinhole leak at 100 atm pressure and 27 °C. The leak continues overnight and is discovered in the morning exactly 12 hours later. The pressure remaining is 30 atm. Assume the ideal gas

law and constant temperature. How many kg of gas have been lost?

Estimate: With T and V constant, the number of moles of O_2 is proportional to the only remaining variable, pressure. If the pressure has dropped from 100 atm to 30 atm., 70% of the gas has been lost, or 70 kg.

Answer: There really is nothing to do here. The 70 kg "estimate" is the answer. This part is trivial.

Part 2: Now suppose the system has a smaller leak, and loses 70 kg, but not simply to the atmosphere. There is also an experiment partially using the oxygen, at a regulated rate of 1.00 moles min^{-1}. This continues for 12 hours, when the leak is finally discovered. What fraction of the of the decrease in mass is due to the leak, and what fraction to the experiment?

Estimate: This looks like we would have to have two simultaneous equations. However, one of the two losses is fixed at a constant value. Let's try the experiment alone first. In 12 hours, which is 720 minutes, the experiment uses 720 moles, at .032 kg min^{-1} (the molecular mass of O_2 is 32 g mol^{-1}) Therefore in 720 minutes the loss is about .032 kg $min^{-1} \times 720$ min ≈ 23 kg . This is of the same order of magnitude as the total, so the losses are comparable. The leak loss is then about 47 kg.

Answer: To two SF, the estimate is the answer. If the data were given to three SF (70.0 kg loss, 12.0 hours) then a little more arithmetic is required, but not much. Roughly 2/3 of the loss is leak.

Part 3: Suppose the loss rate is proportional to the area of the pinhole. Find the ratio of the diameters of the pinholes in part 1 and part 2.

Estimate: The rate of loss in the first case is 70 kg/12 hours, in the second case, 47 kg/12 hours, so the ratio of loss rates is 70/47 = 1.5 to 2 SF. The diameters are in the ratio of the square root of the areas, so $\sqrt{1.5}$, or 1.2 to 2 SF.

Answer: To 2 SF, the estimate is the answer.

Note that each part of this problem is easy. However, suppose it had been presented as "find the ratio of the diameters of the leaks in two oxygen cylinders if each loses 70 kg O_2... with all the details in the initial question. This would have looked like a difficult problem. Breaking it into manageable parts makes it easy.

Part 4: Now suppose that the amount used in both the experiment and the leak is proportional to the pressure in the cylinder. Find the amount used in the experiment and the amount of loss. Assume the initial pressure is the same as in the previous part. Again, the experiment takes 12 hours.

Estimate: The initial rate is known. How long does it take to get to half the pressure? Since only the change in number of moles is responsible for changing the pressure, half the number of moles remains at that point. The relative rates are both proportional to the same pressure, at any given pressure level as the pressure decreases. In part 2, we saw that with the rates fixed, the relative rates were about 1.5. Plot the decrease in pressure against time; The rate $= -CP$, where C is the constant we have to determine, but we know it from the initial condition. If we know some calculus, we could write rate as dP/dt and integrate, so that $\ln(P/P_o) = -Ct$, or $P = P_o \exp(-Ct)\big|_o^{12hours}$. Then find the rate of grams lost at the initial pressure, and as the proportionality constant remains the same, you can finish by getting the final mass; you know the proportionality between mass and pressure from parts 1 and 2. If you don't know calculus, probably the best thing would be to assume the rate remains at the initial value for a short time, then find the amount left after 1 hour, then repeat with the starting rate at the 1 hour level, and determine what is left after 2 hours, and continue for 12 hours. Add the losses over the entire period to get the answer. This is somewhat like a trapezoidal approximation to the integral, although by using the value at the start of each hour, the amount lost will be overestimated. It is not recommended to give a problem like this if the class does not have a calculus prerequisite. However, even without calculus, you should be able to understand this explanation.

6

Stoichiometry

In this chapter, we look at the conversion of one set of stuff to another, and keep track of how much of each we have when we get through. We have immediately one fixed point: whatever the total number of kg we had to start is the number of kg we have when we finish. Matter is neither created nor destroyed. When we have a chemical reaction, some of the starting material is converted into something else; if the starting substances are exactly matched, it is possible to convert all of it, although in most of these problems, a fraction of one or another of the starting materials will be left over, while all of the rest is converted to product. We just have to remember that calculation of what reacts depends on the number of moles that we start with, not the number of kg, or other units (see the introductory example below). So, the first step is always to convert the amount of starting substances, the reactants, to moles of reactants. Then it may turn out that one reactant has more moles than the other can use up (remember that the ratio is given by the stoichiometric coefficients, which need not be in 1:1 ratio). The reagent that will be completely used up, the limiting reagent, limits the amount of product that can be formed, and we use that to find how much product is formed, in moles. The final moles can be multiplied by the molecular mass to get the kg of each product. Again, kg may not be the most convenient unit — chemists are more likely to work with grams, one thousandth of a kg, while chemical engineers may work in tons, with a ton (metric, of course) equal to 1000 kg. Just to give an introductory example, suppose you have 4.03 g of hydrogen and 11.2 L of oxygen at STP. These look like incommensurate units to start, but as soon as we convert to moles, all is well. These react to make water: how much?

For oxygen, we have 0.500 moles of O_2 (by now, you know how to get this, from the previous chapter), and enough data for three SF. For hydrogen, you have 2.00 moles of H_2, meaning 4.00 moles of H (again, three SF). Since the reaction equation is $2\ H_2 + O_2 \rightarrow H_2O$, we see at once that we need two moles of H_2 for each mole of O_2; since we have four moles of H_2 for each mole of O_2, right away we know that we have more H_2 than the O_2 can use, and there must be left over H_2, while the O_2 reacts completely. There is half a mole of O_2 so there is one mole of O atoms, enough to make one mole of H_2O. This is essentially the end of the problem. One mole of H_2O has a mass of 18.0 g (remember, we have three SF in this problem). Let us check: we started with 0.500 mole of O_2, which is 16.0 g, plus two moles of H_2, which is 4.00 g, for a total of 20.0 g of total starting material. We finish with 18.0 g of H_2O plus the left over mole of H_2 which could not react for lack of oxygen; one mole of H_2 means 2.00 g, so the total final mass is 20.0 g, equal to the starting mass, as required. Oxygen is the limiting reagent, hydrogen is the excess reagent. The check is satisfied. Once we know the number of moles, we can convert this not only to grams, but, for example, given a temperature and pressure at which the substance is in the vapor phase, to a volume of gas, or whatever it is that is of interest. Therefore, there are two points to remember: work the problem in moles, and check that the amount of matter you finish with equals the amount you started with. Obviously, if you started with more or less matter than you finished with, you made a mistake. In addition, you must have the same number of moles of each type of *atom* at the beginning and the end; the number of moles of molecules change in reactions, but atoms are not created or destroyed.

SUMMARY: The chapter can almost be summarized in a single sentence: find the moles. All the rest is commentary. Be careful that when you find the number of moles you get a reasonable number — If you have less than a molar mass of a substance you have to have less than a mole of that substance. This seems trivial, but mistakes do get made. Check. In doing the estimate, getting the moles to one significant figure is often quick and easy. The only problems occur when it is not so obvious which is the limiting reagent, in which case it may be necessary to carry out the calculation to perhaps two significant figures; of course, watch out for the stoichiometric coefficients when determining the limiting reagent. There is nothing very profound about stoichiometry. Just keep the bookkeeping straight.

Problems:

1.   The reaction $Pb + PbO_2 + 2\ H_2SO_4 \rightarrow 2\ PbSO_4 + 2\ H_2O$ is the reaction in the lead-acid batteries that have been used in automobiles over the last century. (for lights, etc., not electric vehicles). If 100.0 grams of Pb is used, how many grams of $PbSO_4$ are formed? Atomic mass: Pb, 207.2, S, 32.0, O 16.00, H 1.008

   Estimate: The mass of $PbSO_4$ is obviously greater than that of Pb alone. If we had a full mole of Pb to begin with, we could immediately say 2 moles of $PbSO_4$ are formed; however, we have a little less than half a mole. The answer must be a little less than a full mole of $PbSO_4$. The mass of the sulfate is <100, so there is an upper limit of 300 gm. Since the sulfate is close to 100, the answer should be close to 300 g, but a little less. In other words, 0.5 moles Pb gives 1 mole of $PbSO_4$, so we have about one mole of $PbSO_4$.

   Answer: moles of Pb × 2 × molecular mass of $PbSO_4$ = (100/207.2) × 2 × 303.2 = 292.7 gm.

Here we combined the steps into a single formula. Often it is more convenient to find the number of moles ($2 \times 100.0/207.2$) in an initial step, and then multiply by the product molecular mass, 303.2. This problem is simple enough that doing it all in one step is just as easy.

2.   For the combustion of hexene, $C_6 H_{12} + 19/2 \, O_2 \to 6 \, CO_2 + 7H_2O$

> a)   find the volume of CO_2 produced when 8.00 gm of hexene is burned.
>
> b)   If the temperature is raised to 819°C and the pressure is 12 atm, find the new volume of CO_2.
>
> *Estimate*: In this case, there really is a need to start with the moles, as we are going to go to completely different units — here a gas law problem is loaded onto a stoichiometric problem. The atomic masses can be looked up. Very quickly we can take the 8 gm as almost 0.1 mole, so at STP we have $6 \times 0.1 = 0.6$ moles, so 0.6×22.4, or about 12 L. As to part b), we have multiplied the temperature by 4 on the Kelvin scale [(819+273)/273], the only scale that counts, and multiplied the pressure by 12, so we should have 1/3 the volume, about 4 L. This of course assumes that the CO_2 is an ideal gas, so volume is proportional to T, inversely proportional to P. *Answer*: Just follow the steps in the estimate, but this time use a calculator to get the answer to the appropriate number of SF. Part a), 12.8 L b) 4.27 L

COMMENT: In part b), the pressure is fairly high at 12 atm., so at least the third significant figure would very likely be affected by intermolecular interactions, and we should probably use the van der Waals equation or some other non-ideal gas law for the final answer. As a problem that might occur on homework or an exam, this may be neglected, but if an engineer were to design a plant in which these conditions obtained, it would certainly require more careful analysis.

3.   10.0 g of Zn (atomic mass 65.38) is dropped into 1.00 L of 0.1 M hydrochloric acid, HCl, and reacts according to

$$Zn + 2HCl \rightarrow ZnCl_2 + H_2$$

How many liters of H_2 are produced at 37°C and 1.00 atm? We have not used molarity $\mathbf{M = mol\ L^{-1}}$ before, but we shall use it on practically everything involving a solution from now on. This is worth remembering.

Estimate: Do we have enough HCl to react with all the Zn, or will some be left over? Which will be left over? We see that we need 2 moles HCl for each mole of Zn. We have roughly 0.15 moles of Zn, and 0.1 moles of HCl, not the roughly 0.3 moles we would need to react with the given amount of Zn. HCl is therefore the limiting reagent, and is all used up, with some left over Zn. Since there are 0.1 moles of HCl, we will get 0.05 moles of H_2, as 2 moles HCl are needed for each mole of hydrogen gas (see the stoichiometric coefficients, or just notice that the HCl provides just one hydrogen atom, while the gas requires 2). Under the conditions of the problem, if we take the molar volume to be roughly 25 L at the temperature given, we can estimate a little more than 1 L of gas.

Answer: $V = nRT/p = 0.05 \times 0.0821 \times 310/1.00 = 1.27L$

COMMENT: the problem could have asked how much Zn is left over. Then we would calculate the number of moles of Zn = 10.0/ 65.38 = 0.153 moles. 0.05 moles react, so 0.103 moles remain.

3: Part 1: In one of Lavoisier's experiments, he heated Sn (tin) in air, and found that it gained weight. If we start with 10.00 g Sn, how much weight did the sample gain?

Estimate: Lavoisier did not know about atomic mass, but we do, so we can figure out what must be the answer. The atomic mass of Sn is 118.7. Only the oxygen in the air reacts.

$$Sn + \tfrac{1}{2}\,O_2 \rightarrow SnO$$

We have about 1/12 mole of Sn, so we use 1/24 mole O_2, or about 1.3 gm is added to the weight of the Sn.

Answer: We have 10.00/118.7 = 0.08425 moles of Sn. This reacts with 0.04212 moles O_2, or

0.04212 mol $\times$ 32 g/mol (O_2) = 1.35 g is added to the weight of tin. (Should there be a 4th SF?)

Part 2: Another experiment was done in a sealed flask, volume 10.000 L, again with 10.00 g Sn. The flask is returned to room temperature of 17°C (18th century France was a little chilly in winter). What was the final pressure in the flask? The initial pressure was 1.000 atm. The final mass of tin oxide turns out to be the same as in the previous case.

Estimate: To begin with, about 80% of the air is nitrogen, and can't react, so the minimum possible pressure is 0.8 atm. If all the oxygen reacted, this would be the answer. If there were no reaction, the maximum possible pressure is 1.000 atm. If we guess 0.9 atm, we will necessarily be within 0.1 atm of the correct answer. However, we can go on to get a better estimate. Sn when heated will react completely with oxygen. We have about 1/12 mole of Sn, and 1.35 gm O_2. Now we just need to know how much oxygen was present to begin with. In 10 L at STP we have roughly 0.45 moles total gas, of which 0.09 moles is oxygen. We know from part 1 that 1.3 g = 1.3/32 $\approx$ 0.04 moles, reacts, or about half the oxygen that was present to begin with. It turns out that 0.9 atm is actually a close estimate, although we could not know this at the start. Had we gotten an answer outside the $0.8 < P < 1$ range, we would have known right away that it was wrong

Answer: 1) For nitrogen, the number of moles = PV/RT; The partial pressure is what is relevant here, and for N_2 in air, the mole fraction is 0.78. Then $n = 0.78 \times 1.000 \times 10.0$ L / $.0821 \times 290$ K = 0.3276 moles

2) for oxygen, the initial number of moles = $n(N_2) \times (.21/.78)$ = 0.0882, the (.21/.78) being the ratio of O_2/N_2 in air. 0.04212 reacts, leaving 0.0461 moles. Adding, the total moles = 0.3737. Since we also know the volume and temperature, we can find the pressure.

$$P = NRT/V = 0.3737 \times 0.0821 \times 290/10.000 = 0.890 \text{ atm}$$

By accident our original guess as to boundaries turned out to be very close to right. However, once we had the total amount of oxygen and saw that just about half reacted, the final answer could then be guessed pretty accurately. We wind up with three SF, taking the mole fractions in the gas to be good to three SF.

COMMENT: Lavoisier did not know the ideal gas law; Avogadro was a quarter century in the future. Suppose Lavoisier had used a 3 L flask, so there would be only 0.3×0.084 moles of oxygen. Then all of it would have reacted, the final pressure would be 0.8 atm. and oxygen would have been the limiting reagent. Second comment: the author does not know whether Lavoisier actually did the experiment this way, although reacting tin with oxygen and weighing the result was a crucial experiment for Lavoisier.

4: In the electrolytic decomposition of 27.0 gm of water, how many liters of H_2 are produced at 2.0 atm. and 298 K?

Estimate: To begin with one must, as always, write down the stoichiometric equation:

$$2\,H_2O \rightarrow 2\,H_2 + O_2$$

From this we can see that there is as much hydrogen (moles) as water decomposed (moles). From this, given that water molar mass is 18.0 we have 1.5 moles water, so 1.5 moles H_2. Given the roughly 25 L/mole at STP, we observe the pressure is 2 atm so the molar volume will be half that at 1 atm, we expect approximately 12 L $\times$ 1.5 moles, 18 L, or a little more.

Answer: The 1.5 moles is actually as accurate as we can justify given the number of SF. Then pressure is

$$V = nRT/P = 1.5 \times 0.0821 \times 298/2.0 = 18.3 \text{ L} =$$
$$18 \text{ L to 2 SF}$$

COMMENT: The 2.0 atm is the limiting number of SF. If we had 2.00 atm, the answer would be 18.3 L

5: Lithium peroxide, Li_2O_2, is used as a source of oxygen

$$Li_2O_2 \rightarrow Li_2O + \tfrac{1}{2} O_2$$

Part 1: What is the percent by weight of available oxygen in the Li_2O_2?
Estimate: Oxygen has about twice the atomic mass of lithium, so the mass of oxygen in Li_2O_2 is about 2/3 of the total mass (roughly). Half the oxygen is available, so somewhere in the 30 to 40% range is to be expected.
Answer: percent of available oxygen (one atom) is 16.0/ (32.0 + 13.9) $\times$ 100 = 34.8 %
Part 2: The remaining product of the decomposition of the peroxide, Li_2O, reacts with CO_2 to form Li_2CO_3. Also, when table sugar, sucrose, $C_{12}H_{22}O_{11}$, is consumed (metabolized) by the body it reacts with oxygen:

$$C_{12}H_{22}O_{11} + 12 \, O_2 \rightarrow 12 \, CO_2 + 11 H_2O$$

If Li_2O_2 is used in a spaceship as both oxygen source and carbon dioxide absorber with the Li_2O reaction with CO_2 as the absorber, how much excess CO_2 by weight can be absorbed per kg of Li_2O_2? The CO_2 is produced biologically by metabolizing sucrose.

> *Estimate*: The question almost certainly goes by way of the oxygen; no information is given concerning sucrose, so we can assume there is enough sucrose for it not to be limiting. For each oxygen atom, meaning ½ mole O_2, we produce one mole of Li_2O, therefore we can absorb one mole of CO_2. This would imply the ratio by weight of oxygen to CO_2 is 16 to 44; since the ratio of oxygen to Li_2O_2 is about 1 to 3, there are roughly equal weights of Li_2O_2 to start, and CO_2 absorbed at the end.
>
> *Answer*: Every mole of Li_2O_2 produces one mole of Li_2O, which absorbs one mole of CO_2, so there are $44.0/45.9 = 0.959$ gm CO_2 per gram of Li_2O_2.

COMMENT: This is a surprisingly short answer to what looks like an extremely complex problem. As soon as we notice that the sucrose is irrelevant, we can look for the essence of the problem. For each mole of the lithium peroxide, we can absorb one mole of carbon dioxide. Once we realize this, the rest is easy. We could have complicated the problem by observing that the Li_2O can also react with water to form LiOH. If we did not save the water in some other way, this spaceship would need about twice as much Li_2O_2.

6: Marble is $CaCO_3$. Acid rain is approximately 10^{-4} M H_2SO_4. The acid reacts with $CaCO_3$ to produce CO_2:

$$CaCO_3 + H_2SO_4 \rightarrow CaSO_4 + H_2O + CO_2.$$

2.7 cm acid rain hits a temple with a roof of pure marble, area 137 m^2. How many liters CO_2 at STP are produced by this rainstorm, and how much of the roof is removed — that is, how thick a layer of marble is dissolved by the acid in the rain? The density of marble is 2.9 g cm^{-3}.

> *Estimate*: The molar ratio of $CaCO_3$ to CO_2 is 1:1. We need to know how much rain hits the temple with how much acid, and the amount of $CaCO_3$ per unit layer of the roof. The latter is given by the density and area. We can take this estimate in two steps. 1: How much acid hits the roof? 2: How much CO_2 is produced, as an ideal gas, for this amount of acid?

1) If we have 3 cm = 0.03 m rain × 100 (order of magnitude) m^2, this is 3 m^3 of rain = 3000 L. Then 10^{-4} M H_2SO_4 means about 0.3 moles of H_2SO_4 (3000 L × 10^{-4} mol L^{-1})
2) which produces 0.3 moles CO_2, or 0.3 × 22.4 L ≈ 7 L. The more difficult part is the thickness of roof that is removed. The same number of moles of $CaCO_3$ is removed.

We have the $CaCO_3$ mass given in terms of density, which when multiplied by volume equals mass; we know the molar mass is approximately 100 g, so 0.3 moles is 30 g. The area is 137 m^2 ≈ 100 m^2; the thickness of the layer we want is the volume of roof removed divided by the area. To find the volume use the density: V = 30 g/(2.9 g cm^{-3}) ≈ 10 cm^3 = 0.01 m^3. Then the thickness removed is ≈.01/100 = 1 × 10^{-4} m.

> *Answer*: In this case, we should stick to the estimate, with one SF. Do we really know the amount of rain, and the acid concentration, to two SFs? Here the estimate amounted to nothing more than doing the complete problem to one SF.

COMMENT: We can consider whether the answer is reasonable. Obviously if the rain took off more than about 10^{-4} m the temple would not last all that long. If the thickness of the marble roof were 30 cm. or 0.3 m, there would be only about 1000 rainstorms in the life of the roof, maybe 2000. On the other hand, the builders of the temple centuries ago might never have thought of acid rain, a relatively recent development. If this problem had been given with no temples, just an amount of water with a given quantity of acid and a certain amount of $CaCO_3$, we would have to carry the problem to some appropriate number of SF. In spite of the fact that there are 2 H^+/H_2SO_4, there is only one CO_2/H_2SO_4, and that is what is relevant here.

7: A cube of copper (Cu, atomic mass 63.5) 10 cm on a side, density 8.9 g cm^3, reacts in a 500 L vessel with oxygen at 1000 K, 5 atm. pressure. The reaction is $Cu + \frac{1}{2} O_2 \rightarrow CuO$. What is present when the reaction is complete? Find the limiting reagent and the amount of the excess reagent remaining. Do the problem to 2 SF (Given the data above, is this the right number?). Also, ignore the fact that the Cu density given is at room temperature, not 1000 K.

Estimate: To begin with we have to know how many moles of each reagent are present. There are 1000 cm^3 of Cu, hence 8.9 g $cm^{-3} \times 1000$ $cm^3 = 8900$ g. Therefore 8900 g/63.5 g mol^{-1} gives around 140 moles. In the vessel the number of moles of O_2 is n = PV/RT $\approx 5 \times 500/100 \approx 25$ mol. (Why 100 in the denominator? We rounded off the R = .0821 to .1, good enough for the estimate). It takes 2 Cu to react with an O_2 (we divided the entire equation by 2, compared with all integer coefficients). Obviously, there is not enough oxygen: $2 \times 25 < 140$. We should wind up with around 50 moles of CuO and perhaps 90 moles of Cu unreacted.

Answer: Actually, it turns out that 8900 g /63.5 = 140 moles to 2 SF. Our "estimate" for the moles of Cu was as exact as the problem calls for. A really quick estimate would have rounded the 8900 to 10,000, but in a single step this is just a waste of time. On the other hand, if we got more than about 160 moles, we should realize that we pushed the wrong button on the calculator. As to the oxygen, treating it as an ideal gas, there are 30.5 moles

O$_2$, so the estimate was not horrible, but not exact. Then 61 moles of Cu reacts, leaving 79 behind.

COMMENT on part 1: You may wish to check the mass balance to make sure it works out (but of course it must, if the moles do). Also, this reaction might be awfully slow, if the cube of Cu is really a solid single crystal, then there isn't a lot of surface for the oxygen to attack. What is left behind is also a pretty good vacuum (probably the last few molecules of oxygen wouldn't react, but very little would be left.) Can you think of an application for this?

Part 2: Actually, there would be a complication. There is another compound, Cu$_2$O. As the amount of available oxygen drops, this reaction, which uses half as much oxygen for each Cu atom, becomes significant. Let us try to do this without calculus, using a graphical method — at least, we'll describe a graphical approach, without going through the somewhat lengthy full solution. What determines the relative amounts of the two reactions is the **rate** of reaction. The relative rates, in moles/time, will be *assumed* to be: (rate of Cu + ½ O$_2$ → CuO)/(rate of Cu + ¼ O$_2$ → Cu$_2$O) = P(O$_2$)/3, provided P is measured in atm. As the oxygen is used up, the second reaction, producing Cu$_2$O, becomes relatively more important. The assumption about the rates is arbitrary, and made for the sake of having an interesting problem — rates are an important topic, to be covered later. Here we just postulate rates, without pretending that these are real numbers. We also do not know, from what is given, that the rate of each reaction separately is proportional to the first power of the oxygen pressure; this constitutes a second assumption, and one that is likely to be false. The reaction that uses more oxygen may have a higher power dependence on the pressure of oxygen; we ignore this possibility in this problem.

Estimate: 1) if we know calculus we can do the following:

$$d[CuO]/dt = kp[O_2]$$

$$d[Cu_2O]/dt = (k/3)p[O_2]$$

$$dp[O_2]/dt = -\{k + k/3]p[O_2]$$

We do not need the concentration of copper, as it is a solid and remains constant. We can integrate the third equation to give $p[O_2] = p[O_2]_o \exp(-4k/3\ t)$, plug this into the first two equations, and find the amount formed of each by integrating. Since we have used this problem more to think about such problems than as a numerical example, we will leave it at this. The arbitrary assumptions prevent us from having meaningful numerical estimates; however, if you have parallel reactions in which the values of the constants are known, then this is the way to set up the problem.

2) If we don't know calculus: Here we have to make some sort of approximation because if we are not to do an integral, we would have to plot the rates, or rather their ratio. This is a numerical exercise, and we have no numbers here, so we could only do a plot in relative units. We would start with the oxygen pressure as a function of time, which would quickly show as exponential decay, and then plot $[CuO]$ and $[Cu_2O]$ against time; finally, by taking a number of points and adding the area of the trapezoids that are formed by the initial and final values, get the amount of each — this is another example of the trapezoidal approximation to the integral that we did in the last problem in Chapter 5.

7

Atomic Structure and Spectra

Now we are in a different realm, microscopic rather than macroscopic, looking at the energy levels within an atom or molecule. The word "quantum" is related to quantity, and what got quantum mechanics off the ground was Max Planck's guess that energy on the atomic level might not be continuous, but came in lumps, or quantities. This was a radical break with classical physics, for which energy was always a continuous quantity, and Planck was very reluctant to make the transition. However, if energy is continuous, any amount could be absorbed or emitted, which led to the "ultraviolet catastrophe" — the amount of high energy (ultraviolet) light absorbed or emitted by something with a finite temperature was calculated to be infinite. Obviously, this was nonsense, but it was a clear prediction of classical physics. Something had to change, and Planck correctly guessed that there were separate energy levels, and those with shorter wavelength (ultraviolet) had more energy between levels. Since the product of wavelength times frequency is a velocity, for light waves, with the constant velocity of light, high frequency corresponds to short wavelength. Planck postulated that energy was proportional to frequency, $E = h\nu$, where E is energy, ν is frequency, and h is a constant, now known as Planck's constant. He continued by considering there to be energy levels that were "quantized"; that is, there existed energy levels that were separated by a quantity of energy, and that if the separation were large, there would be high frequency light associated with the transitions between them.

Since high frequency means high energy, and only thermal energy is available to make the transition, the high energy transitions did not occur, and no high frequency light could be emitted. This also applies to the absorption of light. We have already mentioned the connection between temperature and energy; high temperature means high energy. If the temperature is high enough it provides enough energy to make a jump from a lower to a higher energy level, but if the jump is too large, it is not possible to get to the next higher level; high energy states are not populated, so there can be no transitions back to the ground state with emission of energy; no energy at all is emitted from the high energy states. This fixes the classical model, in which at high energy, infinite energy was emitted. With this hypothesis, Planck was able to calculate the observed spectrum within experimental error. We are really concerned with numerical problems here, so this is not the book to go through the theory. That said, we need to acquaint ourselves with some fundamental magnitudes, as well as some underlying relations. To get a sense of the orders of magnitude involved we start with Planck's constant, h; in SI units $h = 6.626 \times 10^{-34}$ J s (J s in fundamental units= kg m^2 s^{-1}). This is a small value; if William Rowan Hamilton, working out the basics of physics in the mid-19th century, had realized that there was such a constant that was different from zero, he probably would have figured out quantum mechanics; at that time, physics meant classical physics, energy was considered continuous, without any need to add a finite quantity of energy to get to a different energy, and the experiments that led to the ultraviolet catastrophe were just beginning. Quantum mechanics began when Planck proposed the existence of such a constant at the beginning of 1901 based on the very accurate new set of data that he got for Christmas, 1900. Einstein used very similar ideas to resolve a problem with the heat capacity of solids in 1905 (his result needed some correction, which was provided by Debye in 1909). Again, classical physics provided a definite, incorrect, prediction for what happened to the heat capacity of solids as temperature approached 0 K, specifically, that this remained constant no matter how low the temperature, all the way to absolute zero; in fact, the heat capacity (the amount of energy absorbed to raise the temperature of a substance) dropped to zero as absolute zero was approached. Again the solution was to quantize the energy levels and realize that when the temperature was too low to make the jumps between levels, there was no energy absorption. There were problems with classical physics that went beyond these, and included the spectra of atoms. These spectra showed clear lines, and, again contrary to classical physics, were not continuous. The spectra of atoms and molecules can be found from quantum mechanics; the first such calculation was done by Neils Bohr for the hydrogen atom only, just before World War 1. After a hiatus for the war, progress resumed in the 1920s and the modern form of quantum theory was created.

We began by saying, in Chapter Two, that atomic nuclei sizes are of the order of 10^{-15} m, while the electron cloud that defines the atom is about five orders of magnitude larger, 10^{-10} m. We typically use two atomic size units, nanometers (nm), 10^{-9} m and Angstroms (Å), 10^{-10} m. In other words, 1 Å is about the size of an atom, more or less. One difficulty with this comment is that the edge of an atom is not well defined; the electron density forms a cloud that trails off fairly sharply, but not exactly abruptly. The cloud is not necessarily spherical either, but some electrons are distributed in a cloud arranged in shapes that are arranged approximately along one or another axis, or may have "nodes" — surfaces where the electron density goes to zero. Calculating these electron clouds is too advanced a topic for an introductory course. The rough sketches in Fig. 7-1 suggest what this means. We can, at this stage, understand that chemical bonds are formed by the exchange of electron density (sometimes whole electrons, as in ionic compounds), and that as a result, the length of a chemical bond is usually in the range of 1–2 Å.

The other order of magnitude that we should keep in mind is that bond energies are typically within an order of magnitude of 10^{-19} J. Not surprisingly, energy units that are not many orders of magnitude from one are defined for convenience, and we will make a list of these, as they are often used in chemistry problems:

For reference: 1 eV $= 1.6022 \times 10^{-19}$ J $= 2.4180 \times 10^{14}$ s^{-1} $=$ 1239.8 nm $= 8066$ cm^{-1} for light

These equal signs equate different units, and should be read to mean that the quantity of energy in each unit corresponds to the rest, with the last three units being for light frequency (s^{-1}) or wavelength (nm). The last energy unit, cm^{-1}, the *wave number*, is an inverse distance, so it is an energy, or frequency, equivalent, not an inverse energy like wavelength. It is defined as the number of wavelengths in the distance used for the unit, here, the cm, and, is the unit typically used in IR spectroscopy. There are other inverse wavelength units possible, but we won't need them. Nanometers (nm) is a wavelength, and therefore an inverse energy. Of course. in doing estimates, we will only use one, or at most, two, figures; these five SF values will be used only in getting the final answers.

A major point that concerns us in this chapter is the interaction of light with matter. Light can provide the packet, or quantity, of energy needed to change the state of an atom or molecule; this is the contrast with the classical picture in which any energy could be absorbed. In this way we can understand that only some frequencies or wavelengths of light or quantities of energy can be absorbed, since light carries energy, and the energy of the light depends on the frequencies or wavelengths. Therefore, we account for the observed spectra showing absorption only at well-defined frequencies; these are different for different molecules. Frequency $\times$ wavelength $=$ velocity for any wave. For light, the velocity of light is a constant, c $= 2.997925 \times 10^{8}$ m s^{-1}; for practically any problem we will do, it might as well be exactly 3×10^{8} m s^{-1}. The 2.997925×10^{8} m s^{-1} value is now actually a defined

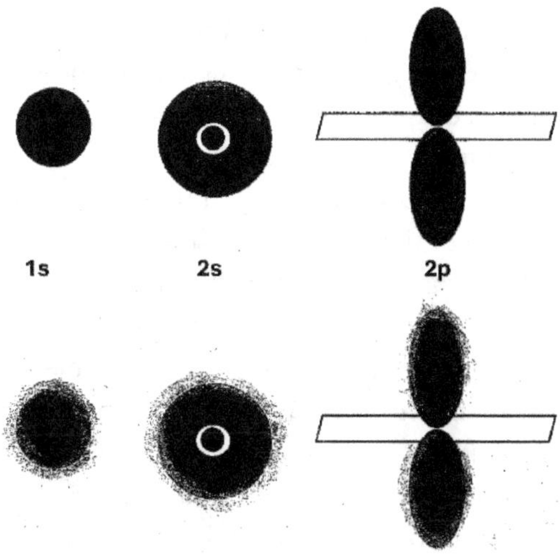

Fig. 7-1.   Rough sketch, not to scale, of the electron density of the three lowest hydrogen atom electron orbitals. The sketches show 1s as closer to the nucleus than the 2s orbital, as well as the nodes of the second shell (2s, 2p) orbitals; the 2s orbital has a spherical node, with the wave function changing sign across the node. The 2p orbital (one of three is shown — there are 3 axes, with a 2p orbital oriented in each of the 3 dimensions of space, each with a unit of angular momentum; the s orbital has no orbital angular momentum) has a planar node, again with a sign change across the node. The lower sketches show the edges as fuzzy; electron densities are probability distributions, which tail off as one gets further from the nucleus, not with sharply defined edges as implied by the upper sketches; sharp edges are sometimes found in textbooks, but this is misleading; what is shown is a single contour of a continuous distribution. For the H atom, the energy of the 2s and 2p orbitals is the same, but for heavier atoms, which have similar orbitals, the 2s energy is generally lower than the 2p energy. For still heavier atoms, the high angular momentum orbitals can have lower energy than the orbitals with the higher numbered "shells" (first quantum number). e.g., Energy[K(4s)] < Energy[K(3d)] — d has 2 units of angular momentum--and it gets more complicated from there on. K (potassium) is only element 19; heavy atoms, with high angular momentum orbitals, need careful calculations and spectroscopy to determine the orbital energies.

value, as part of the SI system; this is part of the present definition of the meter. We use the notation $c = \lambda\nu$, with (Greek letter lambda) λ = wavelength, and (Greek letter nu) ν = frequency. The symbol c is universally used for the velocity of light (it is the c in $E = mc^2$).

λ and ν for wavelength and frequency are also universal symbols. Now we can come back to Planck's constant, and the relation of wavelength and frequency to energy: $\mathbf{E = h\nu = hc/\lambda}$. We will need this relation all through this chapter.

Having established this much, getting estimates requires that we have some idea of the energies of different levels in atoms and molecules. Once again, we choose units close enough to one to make them easy to use. In this case, one of the most used is the electron volt (eV), defined as the energy of one electronic charge dropping through one volt. One electronic charge is 1.60×10^{-19} C and $1\ C \times 1\ V = 1\ J$, so $1.60 \times 10^{-19}\ C \times 1\ V = 1.60 \times 10^{-19}$ J; in other words, $1.60 \times 10^{-19}\ J = 1$ eV (1.6022×10^{-19} J, to five SF). For electronic transitions in atoms, this is also the order of magnitude of a transition (sometimes more, as in the first transition of the hydrogen atom, sometimes less as in the higher transitions, especially of heavier elements, as well as in the higher transitions of hydrogen atoms. How does 1 eV compare to wavelengths or frequencies? If we take 10^{-19} J as E, and round Planck's constant to 10^{-33} J, then, using $E = h\nu$ we get $\nu \approx 10^{14}$ s^{-1}, and $\lambda = c\ /\ \nu \approx 3 \times 10^{-6}$ m, or 3 μm (micrometers). This is in the infrared, longer wavelengths than visible light. Typically electronic transitions for the lighter elements are about ten times higher frequency, and thus shorter wavelength, putting them in the ultraviolet (the visible region is about 0.4 (violet) to 0.7 (red) μm. Since nm are 1000 $\times$ smaller than μm, there are 1000 times as many, and the visible region is then 400 to 700 nm.

While the energy transitions for atoms are in the ultraviolet for the outer electrons, sometimes far into the ultraviolet (the lowest transition for the H atom is close to 129 nm), the inner electrons of heavier atoms can have much higher energy, as the electrons are close to nuclei with much greater charge. Inner electron transitions can be in the X-ray region. This makes physical sense. The larger charge on the nuclei pulls the electrons in even closer, and the energy increases as the distance from the nuclei (positive charges) to the electron (negative charge) gets smaller. Combined with the larger nuclear charge itself, we expect much higher energy, and we get it. We will ignore these transitions from now on because they are not involved in interactions among atoms, hence they are not part of ordinary chemistry.

For molecules, if there are no double bonds, especially if there are only fairly light atoms, as in organic molecules, the first electronic absorptions are in the ultraviolet, wavelength less than 400 nm, but almost always with wavelengths larger than 200 nm for the highest energy transition. When there are double bonds, the wavelength is longer, the energy is lower, as the electrons spread over a larger volume. As we will discuss in the problems, the more the electrons can spread out, the lower the energy. For very large molecules, the absorption can extend to wavelengths larger than 400 nm, and the compound becomes colored, as visible light is absorbed. Chlorophyll and hemoglobin are large and have an arrangement of double bonds that allows electrons to spread over a very large ring in the molecule. They are rather similar molecules, and they are green and red. Many metals have so many levels that many of their compounds absorb in the visible region. Iron rust, for example, looks rusty. Many artist's pigments, like cadmium red, are based on metals. Others have large organic molecules, or various complications, which we won't bother with here.

Unlike atoms, which have only one mode of absorbing light, electronic transitions, molecules can also vibrate and rotate. The vibrations involve the motion of the nuclei, much more massive than electrons, so their motions are effectively "slower", definitely of lower frequency, and thus lower energy. As a result, these produce light absorption in the infrared — as always, longer wavelength goes with lower energy. Small molecules have only a few ways they can vibrate, so they produce only a few wavelengths of light absorption. Larger molecules have many modes of motion, and can produce more complex spectra. Rotations are even slower, longer wavelength, lower frequency, motions, and pure rotation spectra stretch into the microwave region with wavelengths in hundreds of micrometers. We will not consider these further; generally, such matters are dealt with in more advanced courses.

There can also be weak chemical bonds, with lower energy. There can also be non-covalent bonds, of which probably the most important are *hydrogen bonds* like those between the oxygen of one water molecule and hydrogen of a neighboring molecule, which are typically around 20 kJ mol^{-1}, or roughly 3×10^{-20} J for one such bond — in this case there can be a range of over a factor of 2 in bond strength. Hydrogen bonds are responsible for what seems like a remarkably high boiling point for the very small light molecule, water. These bonds are generally several times higher in energy than $k_B T$, the thermal energy at room temperature ($\approx 1/40$ eV), accounting for the water being in a condensed state at this temperature. Similar molecules (e.g, ammonia, NH_3) boil below 0°C. Hydrogen bonds are critical in biology, where they determine the structure of proteins, RNA, and DNA. Other kinds of weak non-covalent bonds also exist, but are less important.

There is a relation between *potentials* and the spacing of energy levels. The relevant potentials are the energies that determine how electrons interact with nuclei, and with each other, either in the same molecule or atom, or with their neighbors. It is sometimes convenient to consider models, not necessarily exact representations of the real system, but simpler approximations that can be calculated easily. The ideal gas was such a model — no real gas is quite ideal, but for many purposes the ideal gas is an adequate approximation. Here we discuss three model potentials (**Fig. 7-2**): that of a particle in a box, in which the energy levels get further and further apart as the energy increases, a harmonic oscillator, where the energy levels are equally spaced, and

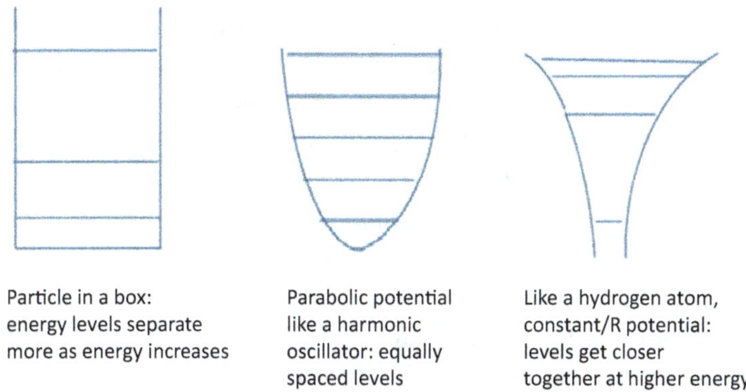

Fig. 7-2.   Three kinds of potentials. Here we see the effect of three shapes of potentials and the consequent distribution of energy levels. (Not to scale).

a hydrogen atom, where a qualitative argument suggests the reason for the levels getting closer and closer together for the upper (high energy) *Rydberg* levels. The upper energy levels correspond to the electron getting further from the nucleus, so the difference in energy from one level to the next gets to be less and less. The potential for the hydrogen atom is actually exact, but it can also serve as a model for more complex systems. Vibrations are best modeled as harmonic oscillators for which the potential is a parabola — that is, the energy of the system increases along a parabola as the amplitude of the motion increases. This is the way a simple spring behaves; such a spring follows Hooke's Law, $F = -kx$ so $E = \tfrac{1}{2} k\, x^2$, where x is the displacement of the spring from its rest position. In other words, the energy is a simple parabola. As usual, classically, energy levels do not exist — any energy is possible. The energy levels in a *quantum harmonic oscillator*, unlike those of a classical spring, are quantized. In fact, the quantum harmonic oscillator is very important in describing real systems, unlike the particle in a box, which does not exactly exist in nature. Harmonic oscillator energy levels can be described by $E_n = (n + 1/2)h\nu$ where h = Planck's constant, ν = the frequency, such that $h\nu$ is the energy separation of the levels, and n is the level. The energy of a transition is $[(n+1+1/2) - (n+1/2)]$ $h\nu = h\nu$. In other words, the separation is always the same.

Where does the ½ $h\nu$ come from? This is a consequence of the *Heisenberg Uncertainty Principle* which states that it is not possible to know two *conjugate quantities* exactly; the position in one direction and the momentum in that direction form such a conjugate pair. If n = 0, the energy is ½ $h\nu$, not zero. If it were zero, the momentum would be zero, and the particle would not have a defined location at all. As it is, the momentum is such that the uncertainty of the position is finite and, although showing this is a more advanced topic, correct. It turns out that the vibrations of molecules can be described by "normal modes", which are like separate oscillators, although they may involve multiple atoms; that is, normal modes have energy levels which are those of harmonic oscillators, each of which is, in principle, independent of the others (not exactly independent in reality). The vibrations of molecules are mostly in the infrared region, with wavelengths generally in the range 1 to 15 μm, or frequencies 3×10^{14} s^{-1} to 2×10^{13} s^{-1}, or energies 2×10^{-19} J to 1.3×10^{-20} J; all these are equivalent, with only the units being different; increasing wavelength corresponds to decreasing frequency or energy, as we saw already.

Units of energy: So far we have used Joules as the unit of energy. We can return to the reference list of energy equivalences that we gave earlier and discuss a little further how they relate. For atomic size magnitudes, Joules are a very large unit on a molecular scale. Units of energy can be in frequency, as in the preceding paragraph. Two units that are commonly used are the electron volt and the wave number. One electron volt (eV) is the energy of dropping one electronic charge through one volt potential, as we saw earlier, 1 eV = 1.6×10^{-19} J, a value small enough that it is convenient for atomic scale energies. The wave number (cm^{-1}) is used for energies in the infrared region, and especially for the energies of infrared transitions as measured in vibrational spectra. 8066 cm^{-1} corresponds to 1.2398 μm, and from this can be converted to all the other energy units. E (J) = hc/λ = 6.626 x $10^{-34} \times 2.9979 \times 10^8/1.2398 \times 10^{-6}$ = 1.602 $\times 10^{-19}$ J. The wave number is the number of waves (cycles) in the distance specified, essentially always cm.

For N atoms, there must be 3N ways to move in three-dimensional space, and this does not change if the atoms are linked to each other in a molecule. The modes of motion are called *degrees of freedom.* Three are needed for the translation of the entire molecule as a rigid body, three for rotation of the entire molecule as a rigid body (two if a linear molecule), so the number of vibrations (normal modes) is *3N – 6,* or -5 if the molecule is linear. For a diatomic molecule (N=2), which is necessarily linear, the number of vibrations is 2×3 -5 = 1. In essence, the diatomic molecule is just like one spring. Taking CO as a typical example, we look up the one frequency. We ignore the apparent doublet that is obtained when one looks at the actual spectrum, and use the standard value 2143 cm^{-1}, which converts to $\lambda = 4666$ nm = 4.666 μm, and other conversions are left as an exercise. For an estimate, look at the ranges given above, and place the value in the appropriate range. Bigger molecules have more complicated spectra, and some of the lines that one might expect to find do not appear experimentally, because certain symmetry rules prevent interaction with light. This is for a more advanced course. Water, which is not linear, with N=3, has 3 normal modes = $3 \times 3 - 6$. However, the vibrational spectrum is complicated, which is an example of how the normal modes are only an approximation.

This largely takes care of what we need to make estimates of the wavelengths of light absorption for atoms and molecules. By using nm for wavelengths (or μm for IR) as our primary unit for estimates, we avoid large exponents. We also realize that the range of electronic and vibrational spectra is somewhat limited, and thus we should be able to recognize when an answer is so far off as to be implausible, thus requiring immediate rechecking. While the relations among wavelength, frequency, and energy are simple and straightforward, we should also have a sense, after a while, of what kinds of atoms or molecules, and which modes of motion, are likely to produce absorption in the infrared, visible, and ultraviolet parts of the spectrum. This is partly dealt with in the particle in a box and harmonic oscillator problems. Even though no physical system is a really a particle in a box, the fact that big boxes mean less energy can be easily seen in large molecules with extended conjugated double bonds. Also, pay attention to the relation of the energy levels for potential curves of the particle in a box compared to those of the harmonic oscillator, to the spacing of levels when the potential energy curve spreads more than the harmonic oscillator potential (that is, the Rydberg levels we just discussed); consider the Coulomb potential, $-q^2/\mathbf{r}$, as in the H atom, as an example of the latter (in three dimensions, use $\mathbf{r}$ for distance instead of x).

In addition to the chunkiness of energy steps, there are additional things about quantum mechanics that seem very strange, when compared to our ordinary experience. In classical physics, one can make measurements of any degree of accuracy, without a problem of one determination interfering with another. In quantum physics, there is a problem when *conjugate variables* are measured; this is the Heisenberg Uncertainty Principle that we just used to understand the ½ h ν as the zero point energy of a harmonic oscillator. In defining conjugate variables, the first thing we notice is that multiplying them together gives the units energy x time. Does this suggest that energy and time are conjugate variables? Yes. The pair we discussed above is perhaps better known: position, and momentum along the same dimension along which position is defined. (units: position in meters, m, momentum kg m s^{-1}, so position × momentum = kg m^2 s^{-1} = energy × time) To see this, we can use Heisenberg's original example. Try to measure the position of a particle with a microscope — in other words, using light. The resolution of the microscope is no better than ¼ the wavelength of the light. That defines Δx, the uncertainty in position. However, as we have seen, light has energy, and thus it has momentum as well; the momentum is h/λ, which is the "kick" the photon gives the particle, and, as it can be anything from 0 to h/λ, $\Delta p = h/\lambda$. Then

$$\Delta x\ \Delta p \approx \lambda/4 \times h/\lambda \approx h/4$$

The Δp is irrelevant if it is perpendicular to Δx. A more elaborate calculation replaces the ¼ with a slightly different factor, but the difference is not enough to worry about. If we try to use shorter and shorter wavelengths to improve spatial resolution, for example, the uncertainty in momentum becomes proportionately larger.

There is a kind of symmetry between particles and waves in quantum mechanics. If the light wave can have a momentum, like a particle does, why shouldn't a particle also have a wavelength? Why indeed? Louis de Broglie proposed that particles had a wavelength $\lambda = p/h$, which gets us back to $p = h/\lambda$. This was experimentally confirmed for electrons by two Americans, Davisson and Germer. Their discovery led to the finding that the electrons in atoms could be described by a wave, with all the consequences that we see in this chapter, and in other chapters as well. It is worth looking at an estimate of the wavelength of an electron with as much momentum as expected from thermal motion at room temperature, about 4×10^{-21} J.

Since $E = p^2/2m$. and the mass of the electron is 9.1×10^{-31} kg $\approx 10^{-30}$ kg, $p \approx 10^{-25}$ kg m s^{-1}, making $\lambda \approx 6 \times 10^{-9}$ m $= 6$ nm, way bigger than an atom. Even a proton with the same energy (at the same temperature, all particles have the same average energy) with 1800 times the mass would have about 0.15 nm wavelength, roughly the size of an atom. Particles do spread out, in quantum mechanics, and the less mass and velocity they have, the more their wavelength. For the moment we won't worry about the consequences, but it is well to keep these numbers in mind.

(note: the unit s^{-1} for frequency is often written as Hz (for Hertz). However, when converting units, it is better to use s^{-1}, since that allows cancellation of units, just as replacing J with kg $m^2 s^{-2}$ does.)

SUMMARY At an atomic or molecular level, energy changes come in chunks. Particles have wavelengths, light has momentum, and the uncertainty principle puts limits on how accurately we can measure. The orders of magnitude of the energy of chemical bonds are useful to keep in mind. We have looked at the relation between the space available to an electron and the separation of the energy levels; the more the electron can spread out, the closer together are the energy levels, and this has consequences for the wavelength at which light is absorbed. Electronic levels in an atom are separated by so much that the lowest transitions of the outer electrons fall in the ultraviolet. As the electrons go farther from the nucleus, which takes more energy, the high energy levels get closer together, so the transitions are lower energy. Small molecules are similar, and the transitions from the "ground state", where the molecule is when not excited, to the lowest excited state, are usually in the ultraviolet. Large molecules with conjugated double bonds in which the electrons find themselves effectively in a big "box", may have transitions in the visible region. The nuclei, being heavier, move slower, with less energy, so their levels are less separated, and the vibrational transitions fall in the infrared.

Problems:

1) What is the de Broglie wavelength of (i) an electron moving with velocity 7.32×10^4 m s^{-1}? Given: the mass of an electron is 9.11×10^{-31} kg (ii) what is the de Broglie wavelength of an automobile weighing 2000 kg traveling with velocity 75 km hr^{-1}?. (iii) discuss the difference between the two cases.

Estimate (i) Electron: de Broglie wavelength $= h/p$, and for a non-relativistic particle (these are non-relativistic, because $c = 3 \times 10^8$ m s^{-1} $\gg 7.32 \times 10^4$) $p = mv$, where $m =$ mass, $v =$ velocity so $p \approx 10^{-30} \times 10^5 = 10^{-25}$, making $\lambda \approx 10^{-8}$m.

(ii)   Car: First of all, convert to appropriate units: we need velocity in m s^{-1}; mass is already in kg. velocity $= 75000$ m/3600 s ≈ 20 m s^{-1}. Therefore momentum $\approx 4 \times 10^4$ kg m s^{-1}, making $\lambda \approx 10^{-30}$ m.

(iii)   There is a difference of about 22 orders of magnitude in the wavelengths of the two cases. The electron wavelength is large compared to an atom, so we know that the atom localizes the electron, or else the velocity given in the problem is unrealistically low. The wave associated with the car is too small to observe by many orders of magnitude. To begin with, this suggests that we have the problem set up right, as this is what we would expect for the relative magnitudes.

Answer: It is left as an exercise to plug in the three figure data to get an "exact" answer — but the exact answer is probably not too meaningful, if the electron velocity is too low, and the car wave unobservable — not even close.

2)   Part 1: Thermal energy is approximately $k_B T$, where k_B is Boltzmann's constant, 1.38×10^{-23} J K^{-1}. What is the de Broglie wavelength of an electron with thermal energy at 298 K?

Estimate: Energy, as usual, equals $\frac{1}{2} mv^2$, and, as $p = mv$, $E = p^2/2m$, so $p = (2mE)^{1/2} \approx 10^{-25}$ kg m s^{-1}, making $\lambda \approx 6$ nm, still huge compared to an atom.

Part 2: Suppose the electron is travelling at 0.1 c (1/10 the speed of light). What is the wavelength in this case?

Estimate: $p \approx 10^{-30}$ kg $\times 3 \times 10^7$ m s^{-1} $= 3 \times 10^{-23}$ kg m s^{-1}, so $\lambda \approx 2 \times 10^{-11} = 0.02$ nm.

COMMENT: this electron has about three orders of magnitude greater velocity, hence momentum, compared to the problem 1 electron, so it has about three orders of magnitude shorter wavelength. We have now covered a range of values of electron momentum, observed that macroscopic objects, like the car, do not have quantum properties that have to be taken into account (not to say they don't exist, but they are of no apparent importance). It might be useful to do the problems over with protons in place of electrons; the mass difference of a factor of about 1800 leads to different results on an atomic scale. This is left for you. Note that we did not need to go past the estimates to get a physical feel for the magnitude of these waves.

3) A photodetector is set up to respond to light of frequency 8.51×10^{14} Hz (Hz = s^{-1}) It records the energy it receives as 6.455×10^{-15} J hr^{-1}. How many photons per second are received from the source on average, over the recording period?

> *Estimate*: Let h $\approx 10^{-33}$ J s, $\nu \approx 10^{15}$ s^{-1} so the energy per photon is about 10^{-18} J. In 3600 s, one has 10^{-14} J, so roughly 10^{-18} J s^{-1}, giving something of the order of 1 photon per second.
>
> *Answer*: Photon energy: $E = h\nu = 6.626 \times 10^{-34} \times 8.51 \times 10^{14} = 5.64 \times 10^{-19}$ J/photon
>
> Received energy per second: $6.455 \times 10^{-15}/3600 = 1.79 \times 10^{-18}$ J s^{-1}
>
> Photons/second = $1.79 \times 10^{-18}/5.64 \times 10^{-19} = 3.17$ photons per second.

This being an average over an extended period, there is no reason for it to be an integer.

COMMENT: The energy was given as around 10^{-15} J, and this took a good deal longer than 1000 s, and each photon was going to be pretty close to 10^{-18} J, so we could have immediately done an estimate of around 1 photon per second. Also, all things considered, this is a pretty weak signal. Maybe some sort of astronomical observation???

4) Solids, especially metals, exhibit a *photelectric effect*; if they are irradiated with light of energy above a minimum energy, called the *work function*, electrons will be ejected; that is, if the light has photons with energy > work function, even if there are relatively few photons. If the photon energy < work function, no electrons will be emitted, even if it is a very intense beam of low energy photons. If the photon energy > work function, an electron will be emitted with energy

$$E = h\nu - \varphi$$

where $h\nu$ is the photon energy, and φ is the work function. In other words, the left over energy after the work function has been used to get the electron out in the first place, goes into electron kinetic energy. (Einstein received his Nobel Prize for explaining this, not for his Theory of Relativity, which was not mentioned in his Nobel Prize citation; he himself regarded this as his most revolutionary paper in 1905, his *annum mirabilis* which included special relativity; by the time got the prize, 1922, he had done general relativity, a much greater achievement.)

The work function for Mg is 3.67 eV (almost always, work function energies are listed in eV). If the surface of a Mg crystal is irradiated with light of 180.3 nm (i) will electrons be ejected? (ii) if so, what will be the kinetic energy of the ejected electrons? (iii) and, if electrons are ejected, what will their de Broglie wavelength be? (iv) what would the frequency be of a photon with energy equal to the kinetic energy of the ejected electron?

Estimate: First of all, we have to get the energy units to be the same. Start with the photon:

1239.8 nm $\approx$ 1200 nm = 1 eV, so this is around 7 eV. (1200/180). Therefore, about half the energy goes into overcoming the work function, and half into the kinetic energy of the electron: h ν > φ and an electron is emitted with energy roughly 3½ eV; then the momentum = $(2mE)^{1/2} \approx (2 \times 10^{-30}$ kg $\times 3.5 \times 1.6 \times 10^{-19}$ J$)^{1/2} \approx 10^{-24}$ kg m s^{-1}. If we compare with previous problems, this looks reasonable. If we found that it took orders of magnitude more or less than the amount with similar conditions, we would know at once that something was wrong — this is another way to use estimates: do you get an order of magnitude that agrees with similar calculations? From this we can get the de Broglie wavelength of $\approx 10^{-33}/10^{-24} = 10^{-9}$ m, or 1 nm, again a reasonable value. Finally, we are asked to get the frequency of a photon with energy equal to that of the emitted electron, 3½ eV = 5.3×10^{-19} J, $\approx 8.4 \times 10^{14}$ Hz.

COMMENT: This completes the estimates, and all we did was essentially do the problem to one or sometimes two SF. Was it worth doing the estimate, since the whole problem could be done, as long as we brought our calculator along? Limiting the arithmetic to what we can do with no calculator is a lot quicker, *and allows us to check that we are getting reasonable orders of magnitude. If we made a mistake setting up the problem, it would be immediately obvious.*

Answer: Since we have essentially done the whole problem, do what we did for the estimate, this time carrying the correct number of SF, so the completed solution is left for the student. Results are: Kinetic Energy of the electron = 3.21 eV; Momentum = 9.31×10^{-25} kg m s^{-1}; λ (de Broglie) = 7.11×10^{-10} m; frequency = 7.76×10^{14} Hz. We see that the actual numerical answers are about 10% off the estimated values, mainly because we rounded φ down about 5%, and then minor round offs did the rest. All in all, the answers are quite reasonable.

INTERLUDE: From what we already have done: we can see some established numerical relations among the quantities that we are working with, specifically correspondence between frequency, energy, and wavelength. The following line, which partially repeats the line earlier in the chapter introduction, might be worth remembering.

Frequency = 5×10^{14} s^{-1} → 6×10^{-7} m (600 nm) wavelength → 3.3×10^{-19} J energy = 2.1 eV

One could go further and remember the corresponding de Broglie wavelength for an electron with an energy corresponding to a 600 nm photon; a 600 nm photon corresponds to a bright yellow verging on orangish. This book is not enthusiastic about memorization, but having a sense of some major quantities is useful — at least to one SF.

5) The hydrogen atom spectrum: One of the earliest indications that quantum mechanics had to be taken seriously was the fact that Neils Bohr in 1911 worked out the spectrum of the hydrogen atom. The way he did it has been superseded by the modern understanding of quantum theory that came along after World War I. However, with only one electron, the simplification Bohr used worked. We do not need to go through either the modern Schrodinger equation or the much simpler Bohr atom calculation. The energy of the transition from level n_1 to level n_2 is given by

$$E = \mathbf{R} \left(1/n_1^2 - 1/n_2^2\right)$$

If $n_1 > n_2$ this corresponds to emission of light; the reverse corresponds to absorption. $\mathbf{R}$ is called the Rydberg constant, and $\mathbf{R} = 2.179 \times 10^{-18}$ J, which is a pretty large energy, equal to 13.6 eV. The transitions from the lowest (ground) level are deep in the ultraviolet. From $n_1 = 1$ to $n_2 = 2$ covers ¾ of the 13.6 eV, way into the ultraviolet. There are transitions that involve level 2 with levels above, with levels 3 and up. Question: find all the levels that are in the visible region.

Estimate: We know the boundaries of the visible region. Starting from the line above, 600 nm is 2.1 eV, so 700 nm, the low energy end of the spectrum, is 2.1* 600/700 = 1.8 eV; the high energy end (blue) 2.1 * 600/400 ≈ 3.1 eV. This means that we must multiply **R** by something between 0.13 and 0.23. That is 0.13 < $(1/n_1^2 - 1/n_2^2)$ < 0.23. If n_1 = 2, the easiest thing is to try n_2 = 3,4,5… to see what fits. Let y = $(1/n_1^2 - 1/n_2^2)$ If n_2 = 3, y = 0.14, just inside the red boundary; if n_2 = 4, y = 0.19, which is probably yellow or green. For n_2 = 5, y = 0.21, bluish but still visible. Actually we could have skipped calculating each level and skipped directly to n_2 = 7, when we finally are at the violet boundary. If the question had asked us to find the wavelength or frequency of each line, then we would have had to go through the entire process — but the question didn't ask that so we could have skipped quickly ahead. As an exercise, work out the wavelengths of each of these lines.

Answer: This is another case where the estimate is the answer, as we need at least 2 SF to even get more or less relevant information. Just to make sure we know how to find the wavelengths, E = $2.179 \times 10^{-18}(1/4 - 1/n_3^2)$ for n_3 = 3 = $2.179 \times 10^{-18} \times 0.1389$ = 3.027×10^{-19} J, which corresponds to **656 nm**.

COMMENT: This series of lines is called the **Balmer series** and is the only series that has energies in the visible region of the spectrum. Transitions to n = 3 from higher levels are even lower in energy, and exist in the infrared region, and we have already seen that n = 1 transitions are in the ultraviolet. The calculation is essentially the same, with only the numbers different. On the surface of the sun, temperatures are roughly 6000 K, about 20 times the value on earth. Thermal energy on earth is around 4×10^{-21} J, so on the surface of the sun it is around 8×10^{-20} J. Since the n =1 to n = 2 transition of the hydrogen atom has energy of ¾ **R** ≈ 1.65×10^{-18} J, even the sun's thermal energy will excite very few of the hydrogen atoms. Calculating the fraction is a more advanced topic.

We can now consider a set of questions that have only qualitative answers, so we do not need our standard Estimates followed by Answers format.

6) Hydrogen-like ions: If the nucleus of a larger atom than H loses all but one electron, we have an ion that is like a hydrogen atom, but with a higher nuclear charge. Will the energy separation be larger or smaller? Explain

   Answer: The larger charge will hold the electron more tightly so more energy will be required to remove it. The larger charge should therefore increase the energy — we expect a larger value of the equivalent of the Rydberg constant. In fact, this increases as Z^2, not just Z, but the qualitative answer is "increases". Part 2 will deal with why this is reasonable. The initial answer is that all the energy levels increase, so the differences, hence the spacings, which are fractions of the Rydberg constant, will increase too.

   Part 2: Why is Z^2 reasonable?

   Answer: Think of the average distance of the electron from nucleus. As the electron is held more tightly, it will be closer to the nucleus. The electron is held more tightly by the charge Z, and the distance to the electron, which also decreases as the charge increases, makes the energy larger. Without looking at the formulas, we cannot say that it is larger by a factor of Z (not Z^n, where n is something other than 1, but n must be positive.). It turns out that n = 1, so overall the factor is Z^2.

   However, this is a discussion of the problem only for hydrogen like atoms; if there is more than one electron, what happens? The outer electron(s) is (are) screened from the nucleus by the inner electrons, which is why the full charge of the nucleus is not what the outer electron feels. There is a smaller effect than the full charge on the energy levels. As a result, a complex calculation is required, well beyond the level of this book. However, we can already understand qualitatively some of the main considerations.

7) What is the lightest hydrogen-like atom to emit an X-ray with wavelength shorter than 1 nm (10^{-9} m). The actual calculation is well beyond the level of this text, so just discuss and make an estimate. "Hydrogen like atom", as in the previous problem, means a heavier atom with all but one electron removed, leaving an ion with a large positive charge.

Answer: For a hydrogen atom, the shortest wavelength comes from energy $(¾)\mathbf{R} = 1.65 \times 10^{-18}$ J $\to$ 120 nm. For a hydrogen-like atom, divide by Z^2 (see the discussion in the previous problem) to find the Z value that gets below 1 nm, which in this case would be $11^2 = 121$, so our first guess has to be Na, atomic number 11.

We can compare to experimental values (*American Institute of Physics Handbook*) in this case,

Experimentally, Mg (atomic number 12) is the first below 1 nm. This is very close to our rough estimate.

COMMENT: Inner shell fluorescence: If an electron is deleted from the innermost shell of an atom, an outer shell electron falls in. If it is from shell 2, then an electron from a higher shell can fall into shell 2. Each such fall must get rid of the difference in energy of the two levels as a form of radiation. It may be a kind of fluorescence, or it may come as electrons. If electrons, this kind of emission is called Auger (pronounced o-ZHAY) radiation. There may be more than

Table 7-1.

Atom	λ (nm)
$_{11}$Na	1.157
$_{12}$Mg	0.989
$_{13}$Al	0.795
^{14}Si	0.6738

one electron, and they may come out as a cascade of radiation.

8) Problem: For this, review Fig. 7-2. We now proceed to some questions that involve different potentials, unlike the electrostatic potential around a nucleus. In Fig. 7-2 we saw the hydrogen atom potential spreading out as it went up. We saw the energy levels getting closer and closer together as they went up. It turns out that when the potential energy is tightly confined by infinite walls (particle in a box), the levels get further and further apart. Question: Can you give any reason for this difference in behavior?

Discussion: We are not asking for a quantitative answer here, so there is nothing to estimate. We observe in the hydrogen atom that the energy spreads out, approaching an almost flat surface at large distances. Flat surfaces correspond to an energy continuum — if there are no constraints, the energy is no longer quantized — any energy is possible. This is close to what happens to a (constant/r) (r = distance) potential when r is large enough. On the contrary, a particle in a box is constrained at all energy values. This still fails to answer the question as posed, but suggests that steep walls might separate energy levels, if a flat surface leads to continuous levels. While not a satisfying answer, short of doing a calculation, it is the best we can manage. It is more nearly a qualitative way to imagine the question.

9) Can we think of an intermediate case, in which the energy levels are equally spaced, getting neither closer together nor further apart with increasing energy?

Estimate and Answer There is no way we can work this out in this book — again, this requires a quantum mechanical calculation--but we can expect a potential that spreads more than the particle in a box, and less than a hyperbola (constant/r) potential. We could try a parabola, which meets these requirements — no guarantee of success, a priori.

However, that does happen to be the answer — it is the harmonic oscillator potential that we discussed earlier. The potential that is associated with a harmonic oscillator is a parabola, and it applies to molecular vibrations as well. In a sense, we could have guessed it. Hooke, in the 17th century proposed the law that bears his name, that the force to stretch a spring is proportional to the extent of the stretch,

$$F = -k_H \, \Delta x$$

where Δx is the extent of stretch, k_H is the Hooke's Law constant, large for a stiff spring, smaller for a weaker spring, and F is the force. We know that force x distance is work, or energy. However, to get the energy required to do this much stretching, we should start with zero stretch, so we are not surprised to find that the energy is less than the final force times the final displacement; it works out as $V = \frac{1}{2} k_H (\Delta x)^2$, with V = potential energy. If you know calculus, you know that you could get this immediately from

$$V = \int_{\Delta x_0} k_H \, \Delta x \, d \, \Delta x = = \frac{1}{2} k_H (\Delta x)^2$$

to get the same formula, mechanically, without much effort. It turns out that the parabolic potential corresponds to normal modes of vibration in a molecule, and that it produces equally spaced energy levels, something that we saw earlier. Proving that the levels are equally spaced, as the question asks, still requires a more advanced calculation.

10) The energy of the nth level of a particle in a one dimensional box is

$$E = (h^2/8 \, m \, \mathbf{a}^2) \, n^2$$

What sort of box might give energy levels about those of an atom? $\mathbf{a}$ is the size — the linear dimension--of the box.
Estimate: Actually, this is just the wrong dependence for the levels of an atom. The difference between levels increases as

the levels get higher: $[(n + 1)^2 - n^2] = 2n + 1$, which obviously gets larger with n. We can however look at two questions: Is there a value for the box size **a** such that the lowest transition is of the right order of magnitude to be an atomic transition? There must be, since we can make **a** anything we want to. What if **a** is about the size of an atom? We start with evaluating $h^2/8\,m \approx 10^{-38}$ kg m^4 s^{-2}. For $h^2/8ma^2$, with **a** about 10^{-10} m, roughly atomic size, then for the lowest $2 \rightarrow 1$ transition, we get something of the order of 3×10^{-18} kg m^2 s^{-2}. The units are correct for energy. Remembering that the lowest transition for hydrogen is about 2×10^{-18} J, this is pretty close. An atomic size box has its lowest transition of the same order as the lowest transition of hydrogen, if the box is the same size. However, the higher transitions are all in the wrong place. The atom is not a particle in a box.

11) So, is there anything in the real world like a particle in a box? It turns out that the energy levels of molecular rotations do have the spacing of $2J + 1$, where J is the rotational quantum number. This is the same kind of spacing as a particle in a box. Think of a reason why this might make sense.

    Discussion. When a molecule rotates rigidly, it comes back to the same position. Its range is bounded in an absolute manner by geometry. In a sense this is like a particle in a box, which has an absolute boundary by the definition of a box. Is this analogy valid? It is not usually thought of. Whether there is a good quantum mechanical reason for considering these to be analogous cases is not so clear. Understanding how it happens that the bounded cases both have the same dependence of energy on quantum number is not a question for first year chemistry, but it is interesting to notice.

12) One class of molecules, the aromatics, has some of its bonding electrons, the π electrons, fairly free within the confines of the molecule. Imagine the electrons to be particles within the molecular "box". The atoms are arranged in rings (see the Figure in this problem, below) for molecular diagrams). Suppose rings are

added to such a molecule to allow it to be twice as large in one dimension. What happens to the transitions?

Estimate/Answer: The different directions of a two dimensional particle in a box are independent, so we can consider one dimension at a time (in a molecule the directions are not independent). Consider light that only excites the electrons along the long direction, and ignore the other two dimensions. Light can be polarized (that is, its vibrations oriented in one plane) so that this is a reasonable experiment. With one dimension doubled (say, $a_x \to 2a_x$) transitions in that direction should have their energies cut by a factor of 4 (the inverse length appears squared in the energy). The reason these molecules are a little like boxes is that the potential is level enough that the electrons can move fairly freely within the molecule, but they hit a potential wall at the edge. We can compare molecules with 1 ring (benzene) to two rings (naphthalene) to four rings (tetracene). Experimentally, these molecules have transitions that change in the expected direction, but not so fast as a particle in a box would suggest. Table 7-problem12 shows that this does exaggerate the analogy to a particle in a box:

COMMENT: This pushes the particle in a box model quite a bit beyond how far it can reasonably be pushed. The

Table 7-problem 12.

Compound	Wavelength (nm)	Energy (eV)
Benzene	254	4.88
Naphthalene	285	4.35
Anthracene	375	3.31
Tetracene	455	2.73
Pentacene	580	2.14

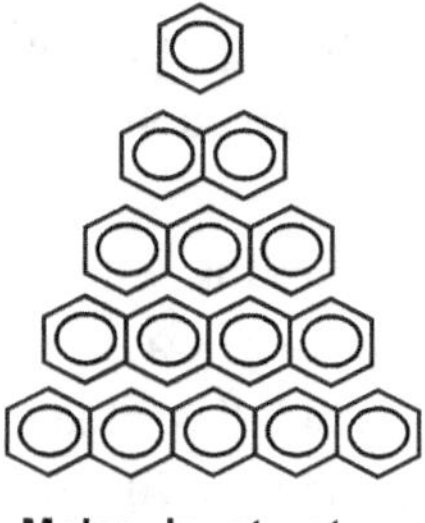

Molecular structure

Fig. 7-problem 12.   The compounds shown here are called "aromatic"; benzene, which has sufficient vapor pressure, actually is. The hexagons show where the carbon atoms are — each corner has a carbon atom. If the corner is one where two rings meet, that is all that is there; if the corner is not one where rings join, there is a hydrogen atom attached, For example, benzene, with one ring, has no joining ring, so its formula is C_6H_6 — every carbon has a hydrogen. Naphthalene, two rings, has two joining corners, so it has the formula $C_{10}H_8$ — two carbons are at joining corners, so have no hydrogens. And so on. The rings inside the hexagons represent electrons that are free (more or less) to move around the rings, and among them. There is a rule (Huckel's rule) that the number of such electrons in an aromatic compound is of the form $4n + 2$; benzene has $4 \times 1 + 2 = 6$, naphthalene has 10, and so on — each ring adds 4 more, to give 14, 18, 22... It almost looks as though these electrons were in a "box" that extends in one dimension as the number of rings increases, but it doesn't quite work that way. Still, we are not surprised at the order of adsorption peaks, which get to be of lower energy as the size of the not-exactly-a-box increases. The last two compounds have absorption in the visible region; pentacene, for example, looks purple.

independence of the two in-plane dimensions is not quite so simple, as the second dimension here is not neglgible — the electron distribution inside the molecule has nodes (positions of zero density that are really two dimensional), for one thing. A truly one-dimensional box also has nodes but they are not the same as the nodes of these molecules. However, some of the behavior of the electrons can reasonably be compared to a one-dimensional particle in a box, to the extent that bigger molecules have lower energy differences between the ground state and the lowest excited state. However, the dependence is less than linear in these molecules, while in a particle in a box, it goes as width squared.

COMMENT 2: Suppose the box became of macroscopic dimensions. Then $1/a^2$ would become so small that the energy level spacing would be much less than thermal energy, or, in effect, continuous. A metal is little like this, and the electron level spacing is continuous in the relevant energy space (in this case the "valence band" and, the "conduction band" — near the band edges everything changes, but let's ignore this). However, the properties at equilibrium, the thermodynamic properties, are far from what would be expected on this simple model, because the electrons obey different statistics — only a fraction of the electrons are free to move, in other words. The particle in a box model is useful for understanding certain fundamental ideas in quantum mechanics, but the world is not made up of "boxes". As the molecules become larger and more complex, we cannot depend on physical intuition gained by extrapolation from simpler molecules.

8

Solids and Unit Cells

Solids are a class of phases characterized by order; their atoms are arranged in repeating patterns all the way to "infinity". This is characteristic of several different kinds of chemical species. There are ionic crystals, such as ordinary table salt, NaCl; there are metals — all metals have at least one solid phase; covalent solids, in which all the atoms are covalently bonded to neighbors, such as diamond. There are crystals of organic compounds that are held in place by intermolecular forces — this includes crystals of molecules like those shown in the previous chapter for a different reason in the Figure in Chapter 7, problem 12. There may be solids with slightly variable composition, like certain high temperature superconductors. On the other hand, glasses do not fit this description, in spite of failure to flow. In glasses, the molecules tend to be large, and to be tangled together, preventing flow, but there is no long range order; in this respect glasses have more in common with liquids than solids. We will not discuss glasses in this chapter, although we might mention them. The problems and calculations will concern real solids, with long range order.

In solids, one can define *unit cells*, repeating units that contain one or more atoms or molecules; when placed side by side in all three dimensions they create the entire crystal. An atom may be shared between unit cells at unit cell faces, edges, or vertices, although when all the fractional atoms are added up, each unit cell will have an integral number of atoms of each type.

The regularity of the lattice that constitutes the solid has another consequence that is of critical experimental importance. It forms a diffraction grating in three dimensions, and diffracts light of appropriate wavelengths. Since the spacing of the atoms is of atomic dimensions, the light must have a wavelength that is comparable to these dimensions. In other words, the light must be X-rays. By examining the angles at which the X-rays diffract, one can work back to determine the crystal structure, including the unit cell size and symmetry, from the diffraction pattern. Historically, this was first applied to simple crystals, and constituted one more early proof that atoms were real, before World War 1. Today, we have computers that can work back to complex crystals, even proteins, and tens of thousands of protein structures have been deduced from their X-ray diffraction patterns. The calculation to get from the diffraction pattern back to the structure is difficult because the diffraction pattern has only half the information, so the calculations required are very extensive. The first protein was worked out around 1960, principally by Max Perutz and John Kendrew, who shared the 1962 Nobel Prize.

A COMMENT ABOUT SOLIDS, AND OTHER THINGS THAT DON'T FLOW.

There can be more than one solid phase of a given substance; water has 19 forms of ice. The ordinary ice which we are familiar with has the molecules arranged in an essentially hexagonal crystal. However, the more the ice is squeezed, the more phases change to denser and denser forms — the pressure squeezes the molecules closer and closer together, on average. These phases are all solids, with proper unit cells (not cubic, but still regular). However, water is also able to produce a glass-like pair of phases. These do not require high pressure, but they do occur at very low temperature. These phases are extremely viscous, and are effectively glasses, defined as amorphous structures that don't flow; unlike crystalline phases, in glasses the atoms are not arranged in a regular lattice. The way the molecules in these phases arrange themselves is still a subject for research.

Polymers, long stringy molecules that are much, much bigger than the small molecules we consider in the problems in this course, tangle together and cannot move. Hence, we get seemingly solid materials like plexiglass or other apparently solid plastics.

This chapter considers the rudiments of how we understand the way atoms are arranged in space. Obviously, crystals are not really infinite, but, as the atoms do not interact when they are separated by a few nanometers, as far as almost all the atoms in the finite crystal are concerned, there is no difference from the surroundings they would have in an infinite crystal. The only exceptions are the atoms very near the surface, a tiny fraction of the total number of atoms in the crystal. When we talk about the unit cells repeating to "infinity" we mean to ignore surfaces. The regular arrangement of a solid's component atoms or molecules is more stable than a liquid or vapor disordered arrangement of these atoms or molecules. As the temperature rises, the disordered

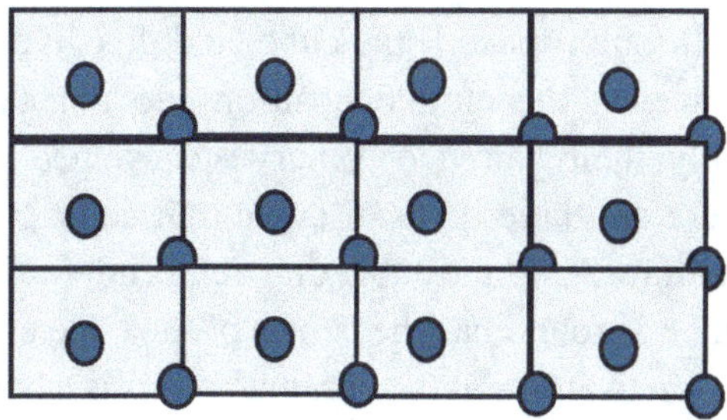

Fig. 8-1. A 2 dimensional crystal with square unit cells containing two atoms, one atom at the center of each cell, and one shared by four cells at each corner (consider only the central cells — surface cells are irregular). Thus each cell has $1 \times 1 + 4 \times \frac{1}{4} = 2$ atoms. The atoms are not shown as occupying the full cell, although they would be in contact with each other — only their positions are shown.

phases are more favored. See the discussion of thermodynamics, Chapter 11.

A two-dimensional example of a unit cell, with 4 neighboring cells sharing square faces, is shown in Fig 8-1. This particular example has unit cells that are square, with the sides of the unit cell being all the same. This example is chosen to have one atom in the center of the cell, and a shared atom at each corner. Rectangles are one of the forms that can be repeated through two-dimensional space. Cubes, the three-dimensional analogues, can be repeated through three dimensional space. There are different symmetries for different unit cell forms. Different types of unit cells can have angles other than $90°$, as long as they fit together by having complementary angles. There exist hexagonal unit cells as well. Since five-fold symmetry cannot be propagated through all space, until fairly recently it was believed that solids with five-fold symmetry were impossible. Like so much else, newer knowledge shows that that which was impossible, is not exactly impossible. This said, we will ignore complications like this, and stick to cubic unit cells, in which all three dimensions have the same length, and all angles are $90°$, for this chapter. Even these cells are not trivial. There is more than one atom per unit cell. What repeats through space is the cube, and this gives the atomic positions. However, the atomic positions do not exhibit cubic symmetry, only their paired or otherwise grouped positions in space do. There are three types of cubic unit cells, two of which are found in nature. A simple cubic unit cell is not found in nature. One form which is found has the center of the cube containing an atom, in addition to atoms at the corners — this is a body centered cell, or *bcc*. Or, one could have atoms in the face of the unit cell, shared with the neighboring unit cell; this kind of unit cell is called face centered cubic, *fcc*. In counting how many atoms there are in a unit cell, count the fraction inside one unit cell. For bcc, there are atoms at the corners, shared with seven other unit cells, 1/8 of each atom in each cell. As there are 8 corners, this means

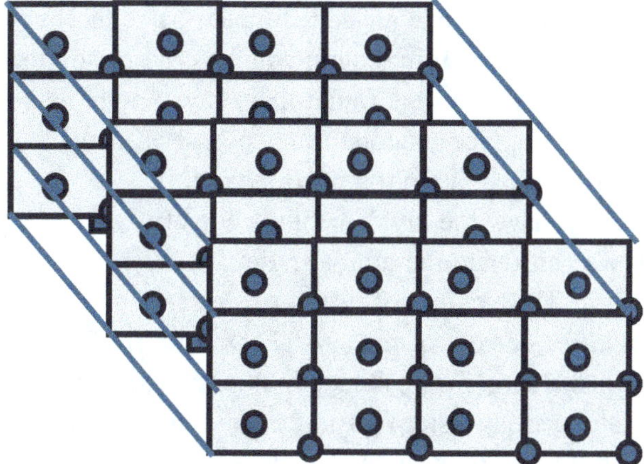

Fig. 8-2:   A schematic representation of a body centered lattice, with one atom in the center of the cell. The cells as shown are not quite cubic, but a rectangular parallelopiped lattice can also be repeated through space. This figure is made by copying Fig 8-1 to project it into 3 dimensions; this time, imagine the corner atoms shared by 8 neighboring cells; again there are 2 atoms/unit cell; the one in the center of the cell, plus 8 × 1/8 for each corner.

the corner atoms add up to one atom. In the bcc, there is one entire atom in the center, so that one has a total of 2 atoms/unit cell. This is the three-dimensional analogue of the two-dimensional case in Fig. 8-1.In fcc, each of the six faces of the cube shares an atom with one neighbor, for a total of three atoms, and, when the 8 × 1/8 atoms at the corners are added, there are 4 atoms/cell. It is convenient to consider a right triangle that defines the geometry of bcc and fcc cubic crystals. The density of atoms in fcc (also sometimes known as cubic close packed) is a little higher than in bcc.

In bcc geometry, the right triangle through the atom in the center of the cube is not isosceles; the three sides are one side of the cube, one face diagonal, and the body diagonal. In the fcc unit cell, the triangle is isosceles; the relevant triangle is set in a face, with two sides and a face diagonal as the three sides of the triangle that defines this type of unit cell. In each case, the hypotenuse equals four atomic radii in length, which is the reason these particular triangles are chosen, rather than, say, a body diagonal for an fcc crystal, or face diagonal on a bcc crystal; neither of the latter has a simple relation to an atomic radius.

Once we have the fundamental geometry of the unit cell defined, we can relate the atomic radius to the length of side of the unit cell. The volume of a unit cell, for a cubic unit cell, is $(side)^3$, and the density of the crystal (D = mass/volume), can be found from the density of a unit cell. For fcc, with four atoms/cell, $D = 4 \times (atomic\ mass/(side)^3)$; for bcc, the same except the prefactor is 2. This does not mean the bcc density is half the fcc density, because the relation between the length of the side of the unit cell and the atomic radius is different. We will return to this, but you can work out the relation between the atomic mass, the atomic radius, and the density, with the information you already have.

To make sure that we have reasonable orders of magnitude, we can try immediately to plug in some numerical values. We expect that the volume should be more than half filled with atoms, so given an atomic radius, we should be able to find the volume of a unit cell, and the volume should be greater than the volume of the atoms in the unit cell, but less than twice that volume. For example, consider an atom with a radius of 1.5 Å, so a volume $(4\pi/3)1.5^3=14.13$ Å^3. Then in 1 cm^3 one has

$$10^{24}/14.13 = 7.07 \times 10^{22} \approx 7 \times 10^{22}\ \text{atomic volumes cm}^{-3}$$

If the unit cell volume is about 40 Å^3, say, and the atomic mass 100 g mol^{-1}, we would have $10^{24}/40 = 2.5 \times 10^{22}$ unit cells cm^{-3}. If this were a bcc lattice, the mass of one unit cell (2 atoms) $= 2 \times 100/N_A = 3.3 \times 10^{-22}$ g, so one gets 3.3×10^{-22} g/unit cell $\times$ 2.5×10^{22} unit cells ≈ 8.3 g cm^3 (Suggestion: go through this set of steps carefully — it will be useful later). If this had been an fcc lattice, 4 atoms/unit cell, while the volume of the unit cell is the same, the density would be twice as great. However, almost certainly the unit cell volume would be larger, so the density would be less than twice as great. Therefore, if the density is known, it might be directly possible to find the difference between bcc and fcc. Why might it not be so easy? Here we assumed that the unit cell volume is given. If it isn't, we have to figure it out, perhaps from an atomic radius. The relation between atomic radius and unit cell volume is different for bcc and fcc. Recall that for bcc, the main diagonal of the cube is the hypotenuse of a right triangle formed by a face diagonal and an edge of the cube, and is itself equal to four atomic radii. The face diagonal is the hypotenuse of an isosceles right triangle with legs equal to the side of the cube.

BCC crystal:(body diagonal)^2= (face diagonal)^2 + (edge)^2, and (face diagonal)^2 equals two (edges)^2, so $(4r)^2 = 3 s^2$, where s = cube edge. Therefore, s = $4r/3^{1/2}$ = 2.31 r, r = atomic radius. The unit cell volume, s^3, is therefore 12.32 r^3, or 6.16 r^3 per atom. For r = 1.5 $\mathring{A}^3$, this means volume per atom = 20.8 $\mathring{A}^3$. The atomic volume, $(4\pi/3)1.5^3$ = 14.14 $\mathring{A}^3$ (carry an extra SF to limit round off error) so the atomic packing fraction, the fraction of space occupied by the atoms, is about 0.680. In this calculation, we actually used the atomic radii twice, and they cancelled; the more economical way to do the problem would be to not get the intermediate values, and keep the calculation in terms of r, or r^3. This result has an additional consequence: the packing fraction in the bcc lattice is always 0.680 for any atomic radius, for a monatomic crystal.

FCC crystal: The packing of fcc crystals is a little tighter, so its atomic packing fraction is a bit larger. There a *face* diagonal is 4 atomic radii. The face diagonal is the hypotenuse of a right isosceles triangle with the sides each one edge of the cube. Therefore, for the fcc we have $(4r)^2 = 2 s^2$, so the side is s = $\sqrt{8}$ r, the volume $8\sqrt{8}\ r^3$. As long as we are doing the intermediate results for r = 1.5 $\mathring{A}$, this makes the unit cell volume 76.37 $\mathring{A}^3$, for four atoms, so we have 19.09 $\mathring{A}^3$ per atom, a little less than in the bcc case, as expected. The packing fraction = 0.740. While, for specificity, we did the arithmetic for a 1.5 Å radius, the packing fraction is the same for any atomic radius. In the problems that follow, using these results will allow you to check your answers — or use them to get the answers in the first place. The problems amount to elementary Euclidean geometry, so they should go quickly.

Other unit cells have specific symmetry arrangements themselves, and there are a finite number of classes of symmetry types, each with specific symmetry related properties. These are not treated in introductory classes, as they require group theory, a mathematical field not normally encountered in the first year.

Summary: Atoms can arrange themselves in ordered or disordered forms. In some arrangements they are disordered, and they are not tightly held together and can flow past each other, as in fluids, which includes liquids and gases. However, some arrangements prevent the atoms from moving through space. In ordered arrangements, the atoms are arranged in unit cells that are distributed uniformly through space, over an "infinite" lattice (obviously not literally infinite). For these, it is possible to calculate the spacing of the atoms, and their density for a given geometry. There are other apparently solid substances that are not crystals, but do not flow, most often because the molecules are entangled, forming glasses; glasses are usually not considered solids, and we restrict the problems in this chapter to crystalline solids. The properties (density, atomic or molecular spacing) can be calculated by geometrical calculations at this level. The symmetry properties are important, but not normally considered in introductory courses.

Problems:

1) Find the dimensions of the fcc unit cell of element X, atomic mass 100, density 8.00 g cm^{-3}.

Estimate: This is a little different than the previous example for bcc, and it has to be done backwards compared to that example — given the density, find the atomic radius (which then gives the side of the unit cube, which is what is asked for). Now there are four atoms/cell, and the volume of a unit cell is (side)^3. Then the mass of a unit cell is $4 \times 100/6 \times 10^{23}$, and mass/volume = 8.00 g cm^{-3}, so

$$\text{side} = V^{1/3} = (\text{density/mass})^{-1/3}$$

For an estimate, we could pick a value in the neighborhood of a few Å for the length of a side, because it essentially always is. There is no calculation needed. If we had gotten a number <1 Å, we could be reasonably sure there was an error — atoms are not that small. In fact, for unit cells of monatomic crystals, anything outside the range of 2.5 – 5 Å for a side of the unit cell is suspect; this may not be true for molecules, which are larger.

Answer: The mass of a unit cell is 4×100 g mol$^{-1}/6 \times 10^{23}$ = 0.667×10^{-21} g.

The density of the unit cell is 8.00 g cm^{-3}, giving volume 0.667×10^{-21} g/8.00 g cm^{-3} = $.0833 \times 10^{-21}$ cm^3 = 83.3×10^{-24} cm^3, so the side of the cubic unit cell = $V^{1/3}$ = 4.37×10^{-8} cm = 4.37 Å

COMMENT: We have used a macroscopic value, 8.00 g cm^{-3} in the same expression with a microscopic quantity; however, in this case we can get away with it, because converting to all microscopic quantities would require division by the same quantity that we would have to multiply by subsequently.

2)   The density of α-iron is 7.865 g cm^{-3}. The atomic mass of Fe is 55.85 g mol^{-1}, and it forms a bcc lattice. How many atoms are in a 10 cm^3 sample of this iron? What are the dimensions of its unit cell?

Estimate: To begin with one expects $< 10^{24}$ atoms of anything in 1 cm^3, so $< 10^{25}$ in 10 cm^3. Actually we know that unit cells tend to be in the 3 to 5 Å range of linear dimensions, so something around 10^{23} atoms in 10 cm^3 seems fairly likely.

Answer: Use the number of atoms per gram, and the density, to get the number of atoms.

$$\text{Atoms cm}^{-3} = 7.865 \text{ g cm}^{-3} \times (6.022 \times 10^{23}\text{atoms}$$
$$\text{mol}^{-1}/55.85 \text{ g mol}^{-1}) = 8.48 \times 10^{22}\text{atoms cm}^{-3}$$

This gives 8.48×10^{23} atoms per 10 cm^3. Then, as bcc has 2 atoms per unit cell, the volume/unit cell is 2 atoms/8.48×10^{22} atoms cm^{-3} = 2.36×10^{-23} cm^3 per unit cell, or 23.6 Å^3 per unit cell.

Then the side of the unit cell = $V^{1/3}$ = 2.87 Å

Note that we probably could have kept 4 SF for this problem, coming out with 2.868 Å. It is arguable that the manipulations would have cost a SF; on the other hand, because of the cube root, it would not be possible with only 3 SF, to go back, take the cube, and get the unit cell volume to the right number of SF, as the round off error increases with higher powers.

3) The NaCl lattice is essentially an fcc lattice, with the center of each face of the unit cell and each corner occupied by Cl$^-$ ions and the center of the cell plus the center of each edge of the cell occupied by Na$^+$ ions, for a total of 4 of each ion per fcc unit cell. There are 6 faces, 8 corners, and 12 edges to a cube. Is this consistent?

Estimate: This is not the kind of question that lends itself to a quantitative estimate. What is needed instead is to determine the fraction of each ion that belongs to a unit cell if it is on a corner, an edge, a face, in the middle, of the cell. Here, looking at Fig. 8-2 is useful.

Answer: At a corner, 8 cells meet, so each gets 1/8 of the atom there; at an edge, four cells meet, so each atom or ion contributes ¼ to each cell. Each face is shared by two cells, so each atom or ion on a face counts as ½ for each cell. The center position is entirely within one cell. For the Na^+ ions, we have 1 center + 12 × ¼ for edges = 4 per unit cell

For the Cl^- ions we have 6 faces × ½ + 8 × 1/8 corners = 4 per unit cell.

Yes, this is consistent. Look at Fig 8-2 to see 8 cells together, and consider the central corner, the edges near the center, the faces of the cells that are adjacent, and the center of the cells.

4)   Silver forms a cubic crystal: Given: atomic mass 107.87, density 10.50 g cm^{-3} atomic radius 1.442 Å. Does it form a bcc or fcc lattice?

> *Estimate*: The question is really asking are there 2 or 4 atoms per unit cell. It is easy to get the atomic mass — obviously just divide the molar mass by Avogadro's number. Mass/volume gives density. We are given the atomic radius, so all we have to do is calculate the volume for each of the two cases, using the radius, and see which gives the correct density.
>
> *Answer*: Molar volume = 107.87 g mol^{-1}/10.50 g cm^{-3} = 10.27 cm^3 mol^{-1}. Note that checking the units in this, and other problems in this chapter (and others) can save a good deal of grief. These units are the same regardless of whether it is bcc or fcc. This problem was largely done as the numerical example just before the problems. Using that result,

fcc: the side of the cube is s = $\sqrt{8}$ × r = 4.078 Å, so volume = 67.85 $Å^3$

Density= mass/Volume = 4 × (107.87 g mol^{-1}/6.022 × 10^{23} atom mol^{-1})/(67.85 × 10^{-24} cm^3)

$$\text{so } D = 10.49 \text{ g } cm^{-3}$$

This is close enough to 10.50 g cm^{-3} to almost certainly be the answer, but if we check for bcc, starting from the earlier result s = 2.31 r = 3.33 Å, for a volume of a unit cell s^3 = 35.96 Å^3. This is for 2 atoms, so

Density = mass/Volume = 2 × (107.87/6.022 × 10^{23})/(35.96 × 10^{-24}) = 9.96 g cm^{-3} (same units for each quantity as above) This differs from the given density by around 5%, outside the range of possible round off error.

10.49/9.96 = 1.053; the packing fractions we found above were in the ratio 740/680 = 1.088. What we are calculating here is not the packing fraction, but it is not surprising to find that it is somewhat similar.

5) The lattice constant for NaCl, s = 5.628 Å, while for NaI s = 6.462 Å; both have the same type of lattice, in which the Na atom is at the center of each edge of the cell, plus a full atom at the center of the cell, while the anion is at each of the eight corners of the cell plus at the center of each face of the cell. Therefore there are 4 atoms of each type in the unit cell. The atomic masses are: Na, 23.0; Cl, 35.5; I, 126.9, gm mol^{-1}

i)   Find the distance between the centers of the two closest Cl$^-$ in a unit cell

ii)   Find the ratio of densities: Density(NaCl)/ Density(NaI)

Estimate: i) Anion distances: We have to find out whether the closest centers are two adjacent faces, two opposite faces, or the corner and center of a given face. The only thing we can do is work out the geometry of the cases. Opposite faces are obviously the same as the cube side. Center to corner of a face is half the diagonal of a face, so 1/√2 of the side, hence closer than the opposite faces case. This leaves the centers of adjacent faces, which is the hypotenuse of a right isosceles triangle with legs s/2, so $(2(s/2)^2)^{1/2}$ = s/√2, just like the center to corner distance.

For Cl⁻ this should be about 4 Å, and for I⁻ a little more

> *Answer*: Now all that remains is to plug in the values: for NaCl, distance = $s/\sqrt{2}$ = 3.98 Å, for I⁻, 4.57 Å. This part required only elementary geometry. Density needs a little more work, but not much. For NaCl, with 4 atoms of each type per unit cell, mass (atomic mass units) = 234.0. Volume = s^3 = 178.3 Å³ Therefore the density in normal units $(234.0/6.022 \times 10^{23})/178.3 \times 10^{-24}$ = 2.17 g cm⁻³. For NaI the density is 3.38 g cm⁻³. (Plugging in the numbers is left to you).

COMMENT: The SF limitation is set by the atomic masses; we could have gone past 3 if we had wanted. However, for all practical purposes, 3 SF is good enough. The numbers make sense, and the known density of NaCl is in fact 2.17 gm cm³ Note that the density goes up with the iodide, but not nearly as much as the mass ratio of NaI/NaCl = 149.6/58.5. While the density does increase, there is an increase in volume that uses up part of the increase in mass.

6) Metal is evaporated onto a 10 cm × 10 cm glass substrate to make a transparent film, which means the film thickness must be <1/4 the wavelength of visible light. The metal has a density of 10.0 gm cm⁻³.

> Part 1: What is the maximum mass of evaporated metal?
>
> *Estimate*: This is going to be a round number problem; the visible region ends at approximately 400 nm wavelength, setting a limit of 100 nm for the film thickness. Again, this is a density/mass/volume kind of problem, but macroscopic this time. We can estimate the dimensions of the film, then find the volume, and finally use the given density to find the mass from the estimated volume. *Doing the estimate*: The limit of 100 nm on the thickness of the film, or 10^{-7} m, or 10^{-5} cm, x the area of

10^2 cm^2, makes the volume about 10^{-3} cm^3. With a density of 10 g cm^{-3} we need about 0.01 g.

Answer: Considering that we have only 1 SF to start with, there is no more to be done. 10^{-2} g is the answer.

Part 2: Part 1 was pretty trivial. Next we want to prepare a thicker film of a different metal. Its thickness is found by measuring interference fringes to give an average 500 nm thickness. 40 mg of metal has been evaporated onto the surface. The metal has an atomic mass of 56.0, and has a bcc unit cell. Can we tell anything about the interatomic distance, possibly including its actual value?

Estimate: We have the volume of the film, and its mass, so we know its density,

$$D = 0.04 \text{ g}/.005 \text{ cm}^3 = 8 \text{ g cm}^{-3}$$

The density, calculated from the unit cell volume and molecular mass, would be

$D = 8$ g cm^{-3} $= 2 \times 56/6.0 \times 10^{23}$ g/V, so V $= (112/6 \times 10^{23})/ 8 \approx 23 \times 10^{-24}$ cm^3, so the side is a little less than 3 Å.

Answer: It is left for you to plug in the numbers; we get 2.84 Å if we carry it out to 3 figures. Once we know the side of a bcc crystal we can get all the geometry. Is it worth it? Also, the side is very small, <3 Å. Maybe some of the data wasn't so good?

COMMENT: How many SF do we really have? Really, as we are given the average thickness to only 1 SF, and the mass to 1 SF, we might as well take the distance to be known to only 1 SF, and the interatomic distances therefore to only 1 SF. Is it worth plugging in to get the "Answer"? No.

SECOND COMMENT. We know how to do the problem, and could get a three SF answer if we had 3 SF to begin with. However, how well do we know whether a thin film is really crystalline? It might be amorphous. Here, it is not so thin, so taking it to be crystalline is reasonable. If the thickness were an order of magnitude less, this would be uncertain. As it is, the side is suspiciously small, <3 Å. Once again, we must think about what the physical conditions are to decide what is a reasonable problem.

9

Solution Calculations

A solution is a homogeneous mixture of two or more substances in a single phase. Typically, we consider a two component solution, with one component present in much larger amount; the component present in large amount is the *solvent,* and that in lesser amount the *solute.* We are almost always concerned with liquid solutions, most commonly aqueous solutions, although many other solvents exist. It is possible to have more than one solute. It is also possible to have two substances in roughly equal amounts, in which the distinction between solute and solvent disappears (for example, ethanol and water form solutions in any proportion). We have to be able to specify the concentration, the amount of solute present in the solution. For this we need a set of units that relate the amount of solute to the amount of solvent. The most convenient for most purposes are 1) *molarity*, designated M, which gives the number of moles of solute in 1 L of solution (first defined in Chapter 6). 2) *molality*, designated m, which gives the number of moles of solute per 1000 g of solvent. The advantage of molarity is that it corresponds to what is easiest to measure, the volume of liquid. Molality however gives the number of moles of solute relative to a fixed number of moles of solvent. With molarity, because the volume of the solution changes as the solute is added, the number of moles of solvent changes, and we do not know the number of moles of solvent as a consequence. This said, it sometimes doesn't matter, especially for dilute aqueous solutions, the kinds of systems we deal with most often. One liter of pure water, weighing 1000 g, has $1000/18 = 55.5$ moles of water, and adding, say, 0.1 mole of some salt won't change the density by enough to matter in most cases (again, we should be careful to consider precision, and the number of SF involved — it might matter in the fourth SF, and this level of precision is sometimes available). However, for more concentrated solutions the difference does matter, and which unit we use will depend on what it is we are trying to learn about the processes in solution.

In what follows, we discuss concentration; however, the functional concentration is often different from the actual concentration, because of the interaction among ions. For extremely dilute solutions, the difference can be calculated, but we will ignore this in this book; it is a more advanced topic. For more concentrated solutions, calculating this effect is very difficult, but computer simulations give something of a useful approximation, and these approximations are continually improving. We include this comment as a mild warning to take some of the numerical values given for solutions here with a grain of salt (for non-native speakers, with a certain sense of caution).

A solute in solution can be diluted by adding more solvent. If we deal with volumes it makes sense to use M, not m. In this case, given that the volume of a solution times the moles/volume gives moles, for any volume, we have

$$M_1 V_1 = M_2 V_2$$

where 1. 2 are the initial and final molarities and volumes. As long as no additional solute is involved, the number of moles of solute is equal to $M_1 V_1 = M_2 V_2$. If solvent is added to make $V_2 > V_1$ then $M_2 < M_1$.

Because there can be more than one solute, it is worth defining one more unit, the *mole fraction*, for which we write X.

$$X_i = \text{moles of } i / \Sigma_i \text{ (moles of all substances i)}$$

The mole fraction of the ith component of the solution is written X_i, and the same notation applies for the solvent. In that case, obviously the sum of the X_i for all components must be 1; the sum in the above equation is taken over all components, including the solvent. Mole fractions are the convenient unit to use in some cases involving multiple components.

Most of the problems we will do involve getting the amounts of solute and solvent correct with various dilutions, and solution preparations. However, there are other questions that come up in practical terms that depend on the amount of total particles per a given quantity of solvent. *Colligative properties* are those that are determined by the total number of solute particles (per liter of solution, as usual, to a pretty good approximation). These include freezing point depression, boiling point elevation, and osmotic pressure. Freezing can be used to separate water from solutes, as the salt or other solute cannot be incorporated into the water crystal lattice; boiling works too, to separate solvent and solute; the water enters the vapor phase differently from the solutes. Reverse osmosis, as described below, is commercially the way to desalinate water. Freezing point depression is important for salting roads to melt ice.

What all these have in common is that the special kind of energy called *free energy* (later we will see that this is related to the quantity *entropy*) of the solvent molecules is lower when the molecules are diluted by solute. It sounds odd to consider the *solvent* to be diluted by the *solute* but that in effect is what happens in solution. In the pure phase the solvent has higher free energy. However, to establish equilibrium, one can lower the temperature at which the solvent is in equilibrium with the pure solvent, or raise its boiling point, or increase the pressure on it. The point is to make the pure solvent and the solvent as it is in the solution have the same free energy to establish equilibrium, at which point molecules can transfer with equal probability from solution to pure solvent, and from pure solvent to solution.

While we will discuss free energy in more detail in Chapter 11 on thermodynamics, we can start now by repeating that the free energy of water is lower in a solution than in pure water — the solute lowers the free energy. Second, the driving force behind a chemical reaction is lowering free energy. Also, if a membrane that passes water but not solute is placed between a solution containing some solute, and pure water, the water will flow from the pure water to the solution, because the pure water has free energy higher than water in the solution. This is essentially the same as dilution of a salt solution when pure water is added. When salt is added to water it is distributed uniformly. It is no surprise that essentially the same thing happens when there is a membrane separating pure water from salt solution, a membrane that allows water to pass, but not solute; the difference is that the dilution can only come by having the water cross the membrane, because the solute cannot do so.

Osmosis: Now that we know that a change in free energy drives chemical processes, we can see how this applies to osmosis; we will repeat what we have just stated. If something is moving, there must be a free energy gradient driving it. If we have pure solvent on one side of the membrane, and solution containing solute on the other side, then the solvent is effectively diluted on the side that contains the solute. It is still the major component, but at less than 100% concentration. This means that it has lower free energy than the pure substance on the other side of the membrane, so there is a driving force to push the solvent from the pure solvent side to the "diluted" side, and the pure solvent thus mixes the solution to be more like the pure solvent. This is the same driving force that causes a solute to spread out uniformly if it is diluted. The most random, or uniform, distribution possible is the stable state.

We can arrange the solution so that the pure solvent rises above the solution (see Figure 9-2). As it rises, a column of solution rises, pushing back down on the membrane; this increase of the pressure adds to the free energy of the solvent in the solution. Eventually, the pressure gets so high that the free energy in the

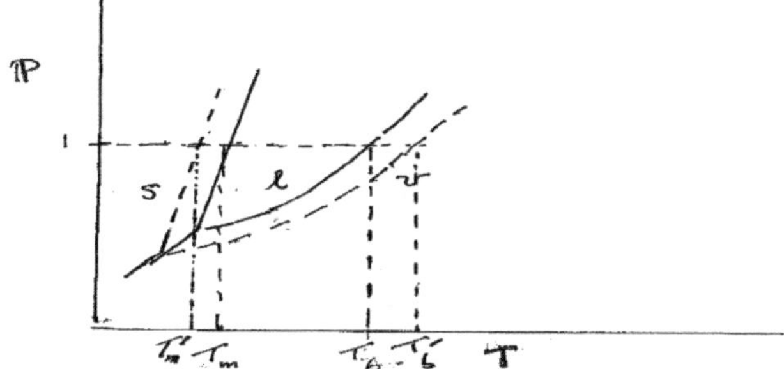

Fig. 9-1. A complex-looking diagram, but really fairly simple. This shows the phases of a typical substance as a function of pressure (vertical axis) and temperature (horizontal axis). The horizontal line labeled 1 on the pressure axis represents 1atm. The solid (s), liquid (l), and vapor (v) phases are labeled; at low temperature, we have solid, at high temperature and low pressure, we have vapor. There is a region between that is liquid. Where the phase boundary crosses 1 atm we have the melting point (the temperature is labeled T_m on the temperature axis). Similarly, the temperature where 1 atm meets the liquid – vapor phase boundary is the boiling point, T_b. We also show what happens when a solute is present; the free energy of the liquid phase drops, giving the dashed lines as the phase boundaries, with the freezing point dropping to T_m', and the boiling point rising to T_b'. This diagram is drawn for a normal liquid, in which the density of the solid is greater than that of the liquid, so that increasing the pressure causes a transition from liquid to solid; the solid — liquid boundary line tilts to the right, because raising pressure at constant temperature causes a phase change from less dense (liquid) to more dense (solid). Water is an exception; the solid is **less** dense than the liquid, increasing pressure still causes a transition from less dense to more dense, but in this case, from solid to liquid, and the boundary line tilts to the left.

solution matches that of the pure water. The difference in free energy of the pure water, and water in the solution, can support a column of water, the height (pressure) depending approximately linearly on concentration of solute in the solution.

Since what counts is the number of particles, and for 0.01 m NaCl one has 0.02 m particles, with the NaCl dissociating into two particles, the osmotic pressure is twice as great (almost — the

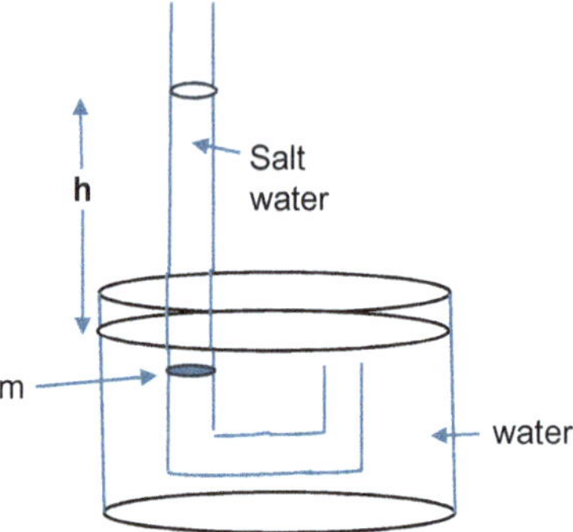

Fig 9-2. Diagram of osmotic pressure apparatus (in principle): There is pure water in the bowl and salt water in the vertical tube. The water in the bowl has higher free energy than that of the salt water. The membrane (m) separates the water from the salt water; it is permeable to water but not to salt. The water moves from the pure water through the membrane to dilute the salt, although it cannot make the free energy the same as that of pure water. The difference is made up by the pressure of the water in the column, which increases the free energy by enough to match that of the pure water in the bowl. The extra energy is mass of water (kg) × gravitational acceleration (m s^{-2}) × the height of the column, h(m). Note: (kg) × (m s^{-2}) × (m) = kg m^{2}s^{-2}, the units of energy. To get pressure, the relevant intensive property, we need the density, not the total mass, of water. The density is mass/volume, so divide by m^3, giving units of kg m s^{-2}/m^2, in other words force/area, the units of pressure. If pressure is added beyond the equilibrium pressure, the water flows *from* the salt water, where, with the added pressure, the water has higher free energy, to the pure water. This is the most common commercial method of purifying salt water, called reverse osmosis.

osmotic pressure will be a little less, because the ions associate with each other to a limited extent, so that they act as effectively being a little less than two full particles). However, this enables us to estimate the osmotic pressure of a solution of different concentrations. Since it depends on molar concentrations, and it is possible to measure values that are very small, it used to be used to estimate molecular weights of very large molecules like proteins — one could measure out a known number of grams, or milligrams, of the substance, determine the molarity from the osmotic pressure, and thus find the number of moles in that mass of protein. If 500 mg/L produced an osmotic pressure rise in the column of water of 1.25 cm = 0.0125 meters, then the molarity was 5×10^{-5} M. Then 0.500 (g/M = $5 \times 10^{-1}/5 \times 10^{-5}$ = 10,000 g/mole. It is very hard to get better than 2 SF this way; determining the column rise to 0.01 cm is barely possible, in a very careful measurement. Now there are better ways to get the molecular mass — the entire structure of the protein can be determined, giving the exact value of the molecular mass — and with the techniques available in 2024, in most cases, the position in space of each atom (recall the discussion in Chapter Eight on X-Ray diffraction of the structure of proteins).

The short form approximation for osmotic pressure, π, is $\pi = \mathbf{MRT}$, with M the molarity, and R, T having the same meaning they had in the ideal gas law. Since M = moles L^{-1}, π has units of pressure, just as it should. This approximation breaks down at higher concentrations, where there are significant ion interactions.

Other colligative properties — *Freezing point depression, boiling point elevation*:

For two other colligative properties, the orders of magnitude are rarely a problem, since the extent of boiling point elevation and freezing point depression are, for reasonably dilute solutions, simply proportional to the molality of particles in solution. Remember to allow for dissociation, as we discussed above. The **van't Hoff factor** by which the molality is multiplied times the particular boiling point elevation constant, and the usually smaller freezing point depression constant to account for the number of particles and their interactions (in sufficiently dilute aqueous solutions, molarity can be used — "sufficiently" depends on ionic charge and the number of SF required). Almost always this gives a number between 1 and 10 times the molality, in Kelvins. The boiling point elevation is harder to measure than the freezing point depression (see discussion on page 121).

The phase diagram for a pure substance, that is, the pressure-temperature plot of phase boundaries, shows where the phase changes occur. If we draw a horizontal line across the pressure – temperature diagram, we can get the temperatures at which the solid-liquid transition, and the liquid vapor transition, occur. In Fig. 9-1 this is shown for the normal 1 atm pressure (read the temperature where the line hits the phase boundary, marked at 1 atm on the diagram, Fig. 9-1) and then at higher temperature, read where the liquid-vapor transition occurs. When the free energy, thus the vapor pressure, of the substance is lowered by adding a solute, one gets the dashed line which is intersected by the one atmosphere horizontal line at a lower temperature for the solid-liquid transition, and at higher temperature for the liquid-vapor transition. Because this is not a full textbook, we will leave it at that, and not describe all of what should be understood about the phase diagram. There are "critical points", where the liquid-vapor phase boundary ends, leaving a single fluid phase, and there can be multiple solid phases (water has as of the present writing 19 known solid phases, with our ordinary ice being only one; perhaps more might yet be found) and all sorts of other complications. At this point we simply note that lowering the free energy of the solvent means about a degree or so lower freezing point and higher boiling point, per unit molality — for some substances, several degrees, but nothing that is orders of magnitude more.

In each case, as long as the solution is not too concentrated, a linear (proportional) response to molality is found. The molality is appropriate, because the number of particles of solvent thus remains constant per mole of solute. The van't Hoff constant must be determined experimentally for each solute, although if there are no interactions among the particles, an ideal case, it can be simply the integer value of the number of particles produced upon dissociation of the solute. The effect on the boiling point and freezing point is a property of the solvent. For water, for example, the increase in boiling point is 0.51 $^{\circ}$C, or K, with concentration in molal units (the constant is formally called the *ebullioscopic constant*):

$$\Delta T = \mathbf{i}\, K_b\, m = \mathbf{i} \times 0.51 \times m.$$

The corresponding freezing point depression constant is 1.86 $^{\circ}$C/molal. $\mathbf{i}$ is the van't Hoff constant.

It does not matter whether we use $^{\circ}$C or K, since we are only interested in the change, and both scales use the same size degree. The van't Hoff factor $\mathbf{i}$ accounts for the fact that there may be more than one particle for a given solute; *e.g.*, for NaCl, there would be two particles, Na^+ and Cl^-, which should give $\mathbf{i} = 2$. In fact, because of the interactions between the ions, experimental values of $\mathbf{i} < 2$ are measured. The difference between the actual value of $\mathbf{i}$ and 2 is a measure of the strength of the interaction of the ions, and in the mid-20th century, a huge amount of effort went into measuring K_b, the ebullioscopic constant, and K_f, the corresponding freezing point depression constant. Because measurements of K_f are more accurate, this was used more. Understanding how the ions in water interact with each other was a major research topic.

Vapor pressure of a solution; Raoult's Law. Suppose we have a non-volatile solute, like sucrose, dissolved in a solvent that has a finite vapor pressure, like water. The vapor pressure of the solution is just the vapor pressure of the volatile solvent, and this is lower than that of the pure solvent because the mole fraction of solvent is less than that of pure solvent — think of the number of molecules of solvent available being reduced by the fraction of molecules that are solvent. Write this as $p_S = X_S\, p_S^{\,o}$, where p_S is the vapor pressure of component S, say the solvent, and $p_S^{\,o}$ is the vapor pressure of the pure component, and X_S is the mole fraction of S. If there is more than one volatile component, the vapor pressure of the entire solution is $\mathbf{P} = \mathbf{X_A p_A^{\,o} + X_B\, p_B^{\,o}} + \ldots$ for as many terms as there are volatile components of the solution. The one non-volatile with one volatile case, like sucrose in water, has only the vapor pressure of the one volatile component. All this said, this applies to an "ideal solution", something that does not exist in nature — for some solutions of very similar compounds this is a pretty good approximation. It assumes that the interactions of molecule A with A, B with B and A with B, are all equal. However, because of molecular interactions between molecules that are not so similar, real solutions have more complications than we can possibly discuss here. We will just use Raoult's Law for simple cases.

There is one more point we must cover: there are multiple ways to find the concentration of a solute in a solution, but we should discuss one in particular, spectrophotometry, in which concentration is determined by the light absorption of the solute. As we saw in the atomic and molecular spectra chapter (Chapter 7), all species absorb light when their energy level differences match the frequency (energy) of the light. We can use the light absorption to give the concentration of the absorbing substance.

We start with Beer's Law, expressed in the form of light transmitted:

$$I/I_o = \exp(-\mathbf{a}\ \mathbf{b}\ \mathbf{c})$$

Where I/I_o = fraction of light transmitted, $\mathbf{c}$ = the concentration, $\mathbf{b}$ = thickness of the cell, $\mathbf{a}$ = "absorption coefficient" for the substance at the particular frequency of the light; taking the log,

$$\ln (I/I_o) = -\mathbf{a}\ \mathbf{b}\ \mathbf{c}.$$

This assumes a definition of $\mathbf{a}$ such that ln rather than $\log_{10}$ is used. However, we can redefine $\mathbf{a}$ such that we use $\log_{10}$, which is often convenient. The quantity $A = \mathbf{a}\ \mathbf{b}\ \mathbf{c}$ is called the absorption. The percent of light transmitted (%T) is e^{-A} or 10^{-A}, depending on whether $\mathbf{a}$ is defined in ln or $\log_{10}$ units. In what follows, we will use $\log_{10}$ units. Then if $A = 1.00$, the percent of light transmitted at the wavelength we are talking about is 10.0%. The most accurate measurements have %T in the $0.25 - 0.6$ range, so the absorption in the $\log_{10}$ units is roughly 0.6 to 0.25.

We know **a** and **b**, and measure I/I_o so we get the concentration **c**. We need to pay attention to units, which are usually: concentration is **c** in moles/L, **b** is in cm, and **a** is in cm^2 $mole^{-1}$. The units essentially cancel, except for a factor of cm^3 $L^{-1} = 1000$. Numerically, the range of $\log_{10}(I/I_o)$ that can be measured with some accuracy is about 0.02 to 2.5 but as just noted, 0.25 to 0.6 is best; above 1 it becomes less accurate, and above 2 it is approaching the range where not much accuracy at all is to be expected. If the cell length b=1 cm, this puts an upper limit of about 2.5×10^{-5} M if **a** is around the maximum possible value of about 10^5 cm^2/mole. Switching to a 1 mm cell allows one more order of magnitude of concentration. However, there are many substances that have smaller values of **a** so that measurements in a 1 cm cell move into a more practical concentration range.

Other ways of determining concentration include measurement of electric conductivity, for solutions of electrolytes — those that dissociate into ions. We won't discuss these in this chapter. However, the colligative properties, especially osmotic pressure, can also be used to determine effective concentrations. If the concentration is known from the way the solution was prepared, the colligative properties and the conductivity give the extent to which the ions associate, which can be complex.

Summary: Solutions have at least two components. We first define concentration units, which enables us to discuss the properties of solutions quantitatively. Then we see the consequences of different concentrations of solutes for the properties of the solutions, including their colligative properties. Ions are hydrated in aqueous solutions, and their hydration can occupy essentially all the water, as the solution concentration approaches saturation. Hydration is discussed in Chapter 12. The colligative properties are one of the ways in which we can determine the extent of interaction of the solute particles. The concentration of the solution can be determined by several methods, one of which is spectrophotometry.

Problems:

1) Part 1: What volume of 0.300 M KCl solution can be made with 10.0 g KCl?

> *Estimate*: 10 g KCl is roughly 0.13 moles. To get up to 0.3 M, one needs less than 1 L; there is not enough KCl for even half a liter at 0.3 M — so there must be less than half a liter — we can estimate roughly 0.4 L.
>
> *Answer*: 10.0/74.6 = 0.134 moles. This gives 0.134 moles/0.300 moles L^{-1} = 447 mL

Part 2: How much more KCl is needed to make 0.500 L of 0.300M KCl? Use the result of Part 1.

> *Estimate*: There are a couple of ways to approach this; probably the easiest is to ask how much KCl is needed to make 0.500 L of 0.300 M. However, since we know how much it takes to make about 87% of our target amount, if we just add about 1/8 as much KCl we should be close. $1/8 \times 10 = 1.25$ g, for a total of 11.25 g; here we might as well not bother to round off — that will happen soon enough.
>
> *Answer*: $(.500/.447) \times 10 = 11.36$ g, but we have only 3 SF, so 11.4 g makes 0.500 L of 0.300 L solution, or 1.4 g more KCl than we had in Part 1.

2) What weight of HCl must be mixed with what volume of pure water to produce 0.300 L solution which is 33.4% HCl by weight? For pure water, density $(4°C) = 1.000$ gm L^{-1}. The density of the solution is 1.175 g L^{-1}.

> *Estimate*: Since we expect the amount of water to be less than the final volume, the upper limit on water would be 300 mL = 300 g since the density is 1 (exact

to as many SF as we need for this problem). If there is roughly 1/3 as much HCl, somewhere in the neighborhood of 100 gm is reasonable.

Answer: $0.300 \text{ L} \times 1.175 \text{ g L}^{-1} = 352.5$ g solution mass, so for HCl, there are $.334 \times 352.5$ gm $= 117.5$ g HCl, and 0.666×352.6 g $= 234.8$ g water, which at density $1.000 \text{ g L}^{-1} = 234.8$ mL

COMMENT: Here the estimate is almost unnecessary, as the procedure to get an answer is essentially the same. However, it is worth looking at the problem quickly to see that the amounts are about 1/3 and 2/3 of what cannot be too far from the somewhat more than 300 g of the total solution. If in going through the calculation, we are far from this, we know we pushed the wrong button on the calculator, or equivalent.

3) 400.0 mL of ethanol (C_2H_5OH) density $= 0.7900$ g mL^{-1} is mixed with 0.6000 mL water, density 1.000, to produce a solution with volume 966.4 mL. Find the mole fractions, molality, and molarity of the ethanol.

   Estimate: To begin with, we need the moles of each. It is probably worth remembering that there are about 55 moles of water in a liter (1000 gm/18 gm mol^{-1}), so here we have roughly 33 moles of water. Similarly, 0.8 g mL$^{-1} \times 400$ mL $= 320$ g ethanol. The molar mass of ethanol, 46, means there are about 7 moles of ethanol, say 7 for estimates. We then estimate the mole fraction of ethanol to be around 1/6 (7/(7 + 33)). To get the molarity we need to extend the number of moles to the amount needed to make 1 L of solution, so multiply by 1000/966.4. For purposes of the estimate, this is about 1, so a lower limit is 7 M. For molal, we take water to be the solvent; the definition of molal is moles per 1000 g solvent. To get to 1000 gm water, multiply moles of ethanol by 1000/600 = 5/3, so roughly 10 m.

Answer: Now we just need the amounts to enough SF. For moles, we have water = 33.303 moles; for ethanol, 400 mL $\times$ 0.7900 g mL^{-1} /46.04 = 6.864 moles

i)   This gives X_{eth} = 6.864/(6.864 + 33.303) = .1709
ii) Molarity: 6.864 $\times$ (1000/966.4) = 7.103 M
iii) Molality: 6.864 $\times$ 1000/600 = 11.44 m

COMMENT: First, the density of water in the last two problems has been set at 1.000 g mL^{-1}. Actually, if we take the density to four places, or even three, we should take account of the temperature. The density is 1.000 gm mL^{-1} at 4°C, which is its maximum density. At 25°C, it is 0.9971 gm mL^{-1} hence using 1.000 is not correct to even 3 SF at 25°C.

The last problem shows a case in which molarity and molality diverge greatly. Note that adding 600 mL of water to 400 mL of ethanol produces less than 1000 mL of solution. The molecules attract each other, and the extended structure of water that produces its low density is partially collapsed, possible because of the loose water structure; however, it is already partially collapsed, and imperfect, in liquid water. In ice, the full structure forms; then the ice is less dense than the water, and icebergs float. This is very unusual, and except for bismuth (which is marginal) no other solid floats on its liquid.

4) A 0.1000 M aqueous NaCl solution has a density of 1.004 g mL^{-1}. Find the (i) weight percent (ii) molality (iii) mole fraction of NaCl

Estimate: We need to find how much water there is. We know that the solution is 0.1 M, so in 1 L there is 0.1 mole of NaCl. The total mass of 1 L is 1004 g, from the density, so we can subtract the mass of NaCl to get the mass of water. As to numerical estimates, dilute aqueous solutions of salts have molarity and molality that do not differ a great deal. Also, however many grams water we have, the moles of water won't be terribly different from the 55.5 moles in pure water, and, knowing that we have 0.1 moles NaCl, we expect mole fraction pretty close to 0.1/55.5, or roughly 10% less than 0.002.

Answer: Consider 1 L of solution. In the 1004 g (1.004 g mL^{-1} × 1000 mL) we have 0.1 moles = 5.385 g NaCl, so the mass of water is 1004 -5 = 999 g (we don't have enough SF to go to fractional grams). Then the number of moles of NaCl in 1000 g solvent is 1000/999 × 0.1000 moles = 0.10010 moles, so the molality is 0.1001 m. The mole fraction, X(NaCl), = 0.1001/(0.1001 + 999/18.01) = 0.001801. Do we really have this number of significant figures

5) An experiment calls for 745 mL of 1.15 M sulfuric acid; you have a 5.00 M solution, so you have to dilute it. What do you need to use to prepare the solution you need?

Estimate: You need to dilute the 5 M solution a little more than 4-fold (4-fold would get to 1.25 M) so you need less than ¼ the number of liters you have to wind up with — somewhere around 0.16 L might be reasonable.

Answer: The number of moles of sulfuric acid that you need is 1.15 moles L^{-1} × 0.745 L, which must equal the number you start from, 5.00 moles L^{-1} × V (L), so V = 0.171 L. This is close enough to our initial estimate of 0.16 L

6) Part 1: 1.437 L of a 0.2364 M NaOH solution is to be neutralized by bubbling HCl through it. What volume of HCl, at STP, is required, if all the gas is absorbed and reacts?

The reaction is HCl + NaOH → NaCl + H$_2$O

Estimate: Again, equate moles. Moles NaOH ≈ < 1.5 × < ¼ ≈ 1/3 mole. We know that the molar volume of any ideal gas at STP is 22.4 L, so somewhere in the vicinity of 7 to 8 L is a reasonable estimate.

Answer: 1.437 L × 0.2364 moles L^{-1} = 0.3397 moles = 7.609 L STP of HCl.

COMMENT: First of all, does this problem make sense? Is HCl a gas at STP? Yes, its boiling point is -85.6 °C. Second, do we really have four SF? The only thing with less than 4 SF that we used is the STP molar volume of an ideal gas. First of all, is this molar volume good to a fourth SF (No). Second, is HCl that ideal? (not to four places). This is a cautionary tale. Doing what seems natural may not be good to 4 SF, which requires a great deal of care in doing experiments, and in considering what goes into calculations. Finally, in this problem we took an acid with one H^+ and a base with one OH^-. If we had, say, H_2SO_4 with two H^+, we would need twice as much NaOH per mole H_2SO_4, Of course, then we would not have a gas law problem, because H_2SO_4 is a heavy viscous liquid at room temperature, more so at STP (T = 0 °C). Also, the reaction really is just $H_3O^+ + OH^- \rightarrow 2\ H_2O$; The Na^+ and Cl^- were present originally as separate ions, and they still are at the end. They are in effect spectator ions.

> Part 2: If a number of moles of $AgNO_3$ equal to the number of moles of NaCl is added, there is a reaction with Cl^- to produce AgCl, which is not soluble and precipitates. How many grams of AgCl are formed?
>
> *Estimate*: Now NO_3^- replaces Cl^- as the spectator ion. We had 0.3397 moles of NaOH, so 0.3397 moles of Cl^-, and therefore 0.3397 moles of Ag^+ added. We will have this number of moles of AgCl.
>
> *Answer*: Molar mass: Ag^+ = 107.87; Cl^- 35.46, so molar mass AgCl = 143.33, and 0.3397 moles gives 48.69 g AgCl.

COMMENT: There is a further reaction: $AgCl + Cl^- \rightarrow AgCl_2^-$, which is soluble. This may matter, at these concentrations, perhaps in the fourth SF. However, if the concentrations were higher, this reaction would become very significant.

7) The vapor pressure of water at 25.5°C is 3.22×10^{-2} atm. What is the vapor pressure of a 1.38 m aqueous solution of glucose ($C_6H_{12}O_6$) at 25.5 °C?

Estimate: We need the mole fraction of water; as glucose is not volatile, it contributes nothing to the vapor pressure. Conveniently, we are given the molality, so we know that there are 1000 gm water, hence, 55.5 moles. 1.38 moles added gives mole fraction around 1/40, so the vapor pressure of the water is likely to be about 39/40 of the vapor pressure of pure water at that temperature.

Answer: With 1.38 moles of glucose, the mole fraction of water is $55.5/(55.5 + 1.38) = 0.976$. Then the vapor pressure $= 0.976 \times 3.22 \times 10^{-2}$ atm. $= 3.14 \times 10^{-2}$ atm.

8) Measuring the osmotic pressure, π, of a solution is, or at least used to be, a standard method for estimating the molecular mass of high mass molecules. Now that structure determination has proceeded as far as it has, this is rarely necessary for proteins. However, for some less well defined polymers, it may still be useful. Let us return to the times of yesteryear, and use osmotic pressure to determine the molecular mass of some type of hemoglobin (different species may produce different hemoglobin, so just saying "hemoglobin" does not automatically define the molecular mass). 550 mg hemoglobin is dissolved to produce 100.0 mL of solution. The observed osmotic pressure is 2.17×10^{-3} atm.

Estimate: We need to find the molecular mass of the hemoglobin; we have the mass of added hemoglobin, so if we find the number of moles, we have the answer. Since we have everything we need to get the molarity from the osmotic pressure, this should be straightforward. From $M = \pi/RT$ and RT around 25, we quickly get something like $M < 10^{-4}$, or, considering that there is 0.100 L, there are $< 10^{-5}$ moles of hemoglobin in the solution made with 0.55 g of hemoglobin. Putting these together, we can estimate somewhere in the vicinity of 5×10^4 g mol^{-1}.

Answer: 0.55 g/(2.17 × 10^{-4}/0.0821 × 298 K) = 6.2 × 10^4g mol^{-1}

As you do this problem, check the units; considering the difficulty of measuring osmotic pressure, 2 SF seems, if anything, generous. 1 atm of osmotic pressure will support a column of water about 10 m high, so 2.17 × 10^{-3} atm supports a column about 2.17 cm ≈2.2 cm high. Trying to read this to 2 SF is a seriously non-trivial task, but possible; three SF seems beyond what we can do.

COMMENT: For polymers that would still be measured with osmotic pressure, there is a mixture of molecular weights, so carrying the calculation to more than 2 SF may have limited meaning, even if it were experimentally possible.

9) Which of the following solutions comes closest to depressing the freezing point of water to -5.0 °C? i) 0.900 m $La(NO_3)_3$ ii) 1.30 m $CaCl_2$ iii) 0.900 m K_2SO_4, iv) 3.40 m $(CH_2OH)_2$ (ethylene glycol). The freezing point depression constant for water is 1.86 °C /molal.

Estimate: K_f is a little less than 1/3 of 5°C/molal, so we need something a little less than 3 m particles. The ethylene glycol is already way over 5 °C depression, even with *i* =1. Which of the others comes closest to providing 3 m particles? Possibilities i) (4 particles/molal × .9 m) and ii) (3 particles/molal × 1.3) are obviously also over, which leaves only the K_2SO_4 solution. Is this close? The van't Hoff factor of 3, with the 0.900 m concentration, is at least close.

Answer: For K_2SO_4, with 3 particles, one would get a 5.3°C freezing point depression, but because the molecular interactions make **i** less than 3, we get just about the 5.0 °C that we need. It is left as an exercise to get the freezing point depression for the others.

10) You have an approximately 7×10^{-5}M solution of methylene blue (MB), which has an absorption coefficient of 8.8×10^4 for a frequency in the red region of the spectrum. To determine the actual concentration of MB, you want to measure the spectrum at that frequency. Would a 1 cm or a 1 mm cell be better?

> *Estimate*: You want an absorption around 1, or a little less, to get an accurate measurement. The absorption coefficient is given with cm^2/mole units. With this concentration and absorption coefficient, a 1 cm cell would have an absorption of about 6, essentially opaque, so there would be no measurement. With a 1 mm cell, the absorption is about 0.6, very much in the range of the most accurate measurement.
>
> *Answer*: This is another case where the estimate is the answer. Use a 1 mm cell. With only two choices it is only necessary to get the right order of magnitude. However, you might as well go on as if the actual absorption were asked for, in a 1 mm cell:
>
> $$A = a\,b\,c = 8.8 \times 10^4 \times 0.1 \times 7 \times 10^{-5} = 0.62$$

where we stop at two SF. If we have a very precise spectrophotometer, it could have one more SF, but this is difficult. We can also get the percent transmission, $10^{-A} = 24\%$ of the light is transmitted. This is enough to get a good measurement.

11) 30.0 mg Fe is dissolved in 100 mL of an acid solution, and oxidized to Fe^{+3}. The absorptivity coefficient is 260 at the measurement wavelength (again, using $\log_{10}$ units). With a 1 cm cell, find the absorptivity and %T.

> *Estimate*: $\mathbf{A} = \mathbf{abc}$. We are given $\mathbf{a}$ and $\mathbf{b}$ and have to find $\mathbf{c}$. 30 mg in 100 mL is 300 mg in 1 L. Since iron weighs a little less than 60 g mol^{-1}, 300 g would be > 5 moles, so 300 mg L^{-1} would be >0.005 moles, or in other words, the solution must be a little more than 5 mM = 0.005 M (mM = millimolar).

Then the absorptivity, $A \approx 260 \times 1 \times 0.300/60 = 1.3$, so the $\%T = 10^{-1.3} = 5\%$.

Answer: replace 60 in the estimate with 55.85, and still more light is absorbed--$\%T = 4.01\%$. In other words, 96% of the light is absorbed. Diluting the solution 2:1 (assuming this is done correctly to three SF) would give a much more accurate result for the light absorption; if we were asked for the amount (molarity, or mass) of the iron, this would be an appropriate thing to do.

10

Chemical Equilibrium with Equilibrium Constants K_{eq}, K_c, K_P

If a chemical reaction is run in a "closed system" (which means no exchange of matter with the surroundings), it will eventually stop reacting. This usually means that it has reached *equilibrium*, and has the lowest *free energy* that it can have; we still have to define free energy; we used free energy in the discussion of osmosis in the last chapter, and here again it refers to the property of the system that drives a chemical reaction. For now, let us just assume that the system, in this case the chemical reaction, has reached equilibrium, lowest free energy, and that nothing will change after this. Then we can write down an *equilibrium constant* for the reaction

$$aA + bB \rightarrow cC + dD$$

as

$$K_{eq} = [C]^c[D]^d/[A]^a\,[B]^b$$

A, B, C, D are the reactants (A, B) and products (C, D) of the reaction, with the quantities in brackets being the concentrations of these reactants and products. In the equilibrium constant, these concentrations are raised to the power of the corresponding stoichiometric coefficients. It is easy to see the reason for these coefficients in the exponents: suppose the reaction is a reverse of

133

dimerization, $A_2 \rightarrow 2\,A$. If we take the concentration of each molecule in the reaction, we would have $K_{eq} = [A][A]/[A_2] = [A]^2/[A_2]$. If we count the molecules and realize that each molecule has to be counted the number of times it appears in the reaction, we get the result above. If we have a pure substance in a reaction (say a solid), its concentration is effectively unity, and it does not appear in the equilibrium constant, and a surrounding solvent that does not react also does not appear in K_{eq}. The value of K_{eq} can be pretty much anything, from huge to miniscule. For that matter, if we consider the reverse reaction, its equilibrium constant is the inverse of the original, so if the original is, say, 10^{10}, then the reverse reaction would equal 10^{-10}. Both mean the same thing; if the forward reaction is $A + B \rightarrow C + D$ the very large equilibrium constant means you get mostly $C + D$; if you write the reverse, $C + D \rightarrow A + B$ the tiny equilibrium constant means that you get mostly $C + D$. Nature, and the universe, do not change just because you reversed the way you wrote the reaction.

K_{eq} is a strong function of temperature, in fact an exponential function.

$$K_{eq} = \exp(-\Delta G^{\circ}/RT)$$

ΔG° is the *change in free energy* in the reaction. For the purposes of this chapter, assume there is a free energy associated with each compound, and the difference between products and reactants determines how far the reaction will go, something that we calculate using the value of K_{eq} that we get from the value of ΔG°. The reason for the superscript $^{\circ}$ on ΔG° is that we need the value at equilibrium; one can define ΔG away from equilibrium, but then the reaction is still continuing; if $\Delta G > \Delta G^{\circ}$, there is still a driving force for the reaction to proceed. In this chapter we are interested in values at equilibrium.

The concentrations for a reaction in solution are normally molar concentrations, that is, expressed as moles L^{-1}. If it is a gas phase reaction, the concentration is given in units of pressure — high pressure is equivalent to high concentration. For a gas phase reaction, we can write K_p, the equilibrium constant in terms of pressure. There is a relation between K_c, the equilibrium constant in terms of concentration, and K_p. For an ideal gas, we can write moles L^{-1} as $n/V = P/RT$, and with n/V as concentration in moles L^{-1}, so $P = cRT$. If we have the pressure in the K_p form of the equilibrium, and there is a change in the number of moles of gas in the reaction, then $K_p = K_c(RT)^{\Delta n}$, where Δn is the change in the number of moles of gas in the reaction. For example, if you have the gas phase reaction

$$2H_2 + O_2 \rightarrow 2H_2O,$$

with all the molecules, including H_2O, in the gas phase, $\Delta n = -1$, so $K_p = K_c(RT)^{-1}$.

The relation given above, $\quad K_{eq} = \exp(-\Delta G^{o}/RT)$
is of critical importance in understanding what is determining the result of the reaction. It is often convenient to use the ln of this relation,

$$\ln K_{eq} = -\Delta G^{o}/RT$$

it turns out that the ΔG^{o} used with K_p or the ΔG^{o} for K_c gives slightly different values for ΔG^{o}, because the values for the compounds in the reaction refer to different *standard states*. This is a subtlety that we will ignore for the moment, but if you look up the values of free energy, or the equilibrium constants, you should be aware of this point. This comes up in this chapter because we need it, but we reserve a full explanation for Chapter 11.

Each compound has a free energy of formation from its elements under standard conditions, and a general reaction gives the difference in free energy of formation, (products – reactants), in the course of the reaction. There is a special set of reactions, in which the compound is formed from its elements. The elements are taken to have zero *free energy of formation* from their elements (themselves), usually in whichever phase they are in at 298 K and 1 bar pressure as the *standard state*. For gases, this normally means a hypothetical state in which the gas is ideal at

1 atmosphere pressure.. There are no real gases that are actually ideal. We will ignore this distinction, but in more advanced courses, it will be necessary to take it into account. For example,

$$H_{2(v)} + \tfrac{1}{2}\,O_{2(v)} \rightarrow H_2O_{(l)}$$

is the reaction in which one mole of water is formed in the liquid phase, at one bar, 298 K, from one mole of hydrogen in the vapor phase at one bar, 298 K plus one half mole oxygen in the vapor phase at one bar, 298 K. The ΔG° of this reaction is the free energy of formation of water in its standard state. Suppose water is then a reactant in another reaction, say

$$H_2O + \tfrac{1}{2}\,O_2 \rightarrow H_2O_2$$

ΔG° for this reaction at one bar, 298 K is **ΔG°(formation, H_2O_2) – ΔG°(formation, H_2O).** ΔG°(formation, H_2O_2) is the ΔG° of the reaction $H_2 + O_2 \rightarrow H_2O_2$ at the H_2O_2 standard state.

Obviously there is a consistency requirement among these reactions that may be useful in checking the answers to problems.

The rate of change of **ΔG°** with temperature gives two other thermodynamic quantities; again, we wait a chapter for a discussion of this. Once we have the values of **ΔG°** we have the value of K_{eq}, so we can use it in estimates. It is useful to keep in mind the value of RT, about 2.5 kJ near 298 K. (R = 8.31 J K^{-1}, RT = 2.48 kJ at 298 K). To get some general sense of how this helps us in making estimates, we start with comparing RT with reasonable **ΔG°** values. Chemical bond energies can be much larger than RT, and the difference in energy in the reactants and products can easily be much larger. Chemical reactions most often require that bonds be broken, and new bonds formed. If ΔG° is much larger than RT and negative, then there is almost all product; for example suppose the reaction **ΔG°** is –20 kJ, not a very large value; then $(\exp(-(-20/2.5))) = e^8 = 3000$; with a K_{eq} that large, the estimate is that there is mostly, indeed nearly all, product — and this is with a relatively small **ΔG°**. If **ΔG°** is *positive* and much larger than RT, there is almost all reactant; the reaction hardly begins. Only if there is a near equality in bond strength of reactants and products, so that ΔG^o is of the order of RT in magnitude, do we have much of a calculation to worry about.

Summary: The driving force for a chemical reaction is the difference in free energy between reactants and products. If this difference is comparable to thermal energy, the equilibrium between reactants and products will allow finite amounts of both to exist. If the absolute magnitude is much more than thermal energy, the lower free energy determines where the equilibrium will lie; it will be on the side of the equation with the lower free energy, whether this is reactants or products.

Problems:

1) Consider the vapor phase reaction $PCl_5 \rightarrow PCl_3 + Cl_2$. A fraction, α, dissociates. In the system, the initial pressure is constant at 1 atm and α is measured with the following results (Table 10.1). Find K_p at 450, 500, and 550 K, and the corresponding values of ΔG^o.

   Estimate: Since the reaction goes from having less than half dissociated near the lower end of the temperature

Table 10.1

T(K)	α
400	0.050
425	0.109
450	0.191
475	0.286
500	0.393
525	0.500
550	0.599
575	0.682
600	0.753

range to more than half dissociated near the upper end, we should expect ΔG^o to be roughly in the range of RT in this temperature range. At 500 K, RT is about 4 kJ (8.3 × 500). We observed to start that at the low end of the temperature range the reaction has $\alpha < 0.5$, while at the upper end of the temperature range $\alpha > 0.5$, which is also consistent with a value of $|\Delta G^o| < 4$ kJ near the middle of the range. There it may change sign, as we go from mostly not dissociated to mostly dissociated at high temperature. There is nothing that says ΔG^o is constant with respect to temperature; in fact, the slope of the ΔG^o vs. T curve gives two other thermodynamic quantities, as discussed in the next chapter. From the degree of dissociation at 450 K we expect $\Delta G^o > 0$, and from the value at 550 K, $\Delta G^o < 0$; in both cases by not much more than 4 kJ mol^{-1} in magnitude (If $1/2.718 < K_p < 2.718$, then $|\Delta G^o| < RT$). Then K_p should be near 1, a little less at 450K, closer at 550 K.

Answer: If we start with p = 1 atm of PCl_5 and pressure = 0 of the products, then the concentrations (here, pressure) after dissociation are: $p(PCl_5)$= 1- α, $p(PCl_3)$= $p(Cl_2)$ = α.

Then $K_p = \alpha^2/(1 - \alpha)$ for 1 atm initial pressure

Plugging in values from Table 10.1: T = 450 K K_p = 0.045
500 K K_p = 0.254
550 K K_p = 0.894

By the time we get to 600 K, K_p = 2.29 > 1.

To find ΔG^o, use $\Delta G^o = - RT \, ln \, K_p$.

At   450 K -8.31 x 450 *ln* 0.045 = 11.6 kJ
500 K -8.31 x 500 *ln* 0.254 = 5.7 kJ
550 K -8.31 x 550 *ln* 0.599 = 0.5 kJ

Table 10.2

T (K)	α	K_p	ΔG° (kJ)
462	0.244	0.0636	+10.6
485	0.431	0.244	+5.69
534	0.745	1.99	–3.06
556	0.857	4.99	–7.42
574	0.916	9.32	–10.6

COMMENT: To begin with, the values in Table 10.1 are not quite real — they are generated by ChatGPT — but they are fairly close to real.

Suppose you wanted the values at, say, 525 K. How might you get these, if all you have is Table 10.1? You could try interpolating. Should you interpolate on ΔG°? on K_p? It can't be both, as the relation between the two is non-linear. Looking at the results at the three temperatures at which we found K_p, it can't be K_p, which is obviously far from linear. However, the differences over 50^0 intervals below and above 500 K for ΔG° are 5.9 kJ and 5.2 kJ, close enough to consider interpolation, taking the ΔG° vs T curve to have almost constant slope. This slope tells us a good deal about the reaction — see the following discussion of enthalpy and entropy, limited though it is because we don't use calculus. This would give us ΔG°(525 K) ≈ 2.5 kJ, and we should probably allow ± 0.3 kJ. Possibly we could do a little better by plotting the ΔG° values, and fitting a curve, but that goes beyond what we need to understand here.

2) A liquid, Z, with a vapor pressure of 38 Torr at temperature T (760 Torr = 1 atm, so 38 Torr =0.050 atm) reacts in a sealed vessel with a gas X, present originally at pressure 0.1 atm. as follows: $X_{(g)} + Z_{liq} \rightarrow A_{(g)} + U_{(s)}$ The solid product, U, is soluble in liquid Z. X and A are practically insoluble in Z. K_p for the reaction = 0.01. Find the final pressure in the vapor phase

if Z is present in large excess, so it is not used up. The volume of the gas phase is 199 mL. Also, find the final partial pressure of A. The reaction occurs in the gas phase.

Estimate: We begin with two observations: $K_p \ll 1$, so there is not much reaction, and there is remaining Z as a liquid with a vapor pressure of 38 Torr at the end, thereby setting a lower limit for the final total pressure. (A word of caution here — by dissolving some U, the vapor pressure will drop a little — see the discussion of colligative properties in the last chapter. We assume this effect is negligible here — why is this reasonable?). The next step is to find an upper bound. This is trivial, as there is 0.1 atm = 76 Torr of X, and very little reacts, so you still have 114 Torr total (38 Z + 76 X). This is also the final answer, as every molecule of X that is used up produces a molecule of A, also a gas, so the number of gas molecules is unchanged in the reaction. Within the ideal gas approximation nothing has changed.

Answer: Even though we know nothing about the reactants or products, we can still get a reasonable approximation from K_p, which is given.

$$K_p = p_A/p_X p_Z$$

with the reaction happening in the vapor phase — Z reacts in the vapor phase, not at the interface of the liquid and vapor phase. The partial pressures are $P_X = 0.1 - p_A$, and $p_Z = 0.05$ atm. Then

$$K_P = p_A/(0.1 - p_A)0.050 = 0.01$$

Solving for p_A we get $p_A = 5 \times 10^{-5}$ atm = 0.038 Torr. In other words, not much reaction has happened — of course, we knew this to begin with.

COMMENT: We solved with the approximation that $p_A \ll p_X$ which was justified because $K_p \ll 1$. If $K_p \approx 1$, this assumption breaks down, and we must solve the K_p equation as a full quadratic equation, giving $p_A \approx 3.6$ Torr. ≈ 0.005 atm $\approx 0.1 p_X$. The approximation $p_A \ll p_X$ is not even good to 2 SF. If a third SF were required, the approximation is no good at all. For $K_p > 1$, the approximation is clearly no good. This problem is a good example of the need to pay attention to the magnitude of the values involved, rather than blindly plug into formulas.

3)   An important reaction in air pollution is $SO_2 + \frac{1}{2} O_2 \rightarrow SO_3$

The ΔG^{o} for formation of $SO_2 = -71.79$ kJ ol^{-1}; of SO_3 -88.52 kJ mol^{-1}.

Suppose 200 metric tons (1 metric ton = 1000 kg) are dumped into an atmospheric inversion over a town, effectively confining it to a volume of 50 km^3. *Find the final concentration of SO_3.* Assume air is an ideal gas. SO_3 is the acid anhydride of sulfuric acid (H_2SO_4 –just add water to get the acid). Concentrations of SO_3 of roughly 5 ppm (parts per million) are lethal — are the people in the town in danger?

Estimate: To begin with, this is a one figure problem. Estimating the size of the bowl under the inversion can't be very accurate, for one thing. For estimating the concentration, even before we begin, we can see what we have to do. As with concentration problems generally, start with finding the number of moles. We have 200 ton $\times$ 1000 kg ton^{-1} $\times$ 1000 g kg^{-1} $\times$ 1/64 mole g^{-1} $\approx 3 \times 10^6$ moles. The volume is 50 km^3 = 50×10^9 m^3 $\times$ 10^3 L m^{-3} $= 5 \times 10^{13}$ L. Since 3×10^6 moles would occupy around 8 $\times 10^7$ L, at 25 L mol^{-1}, you have around 8×10^7 L/3×10^{13} L $\approx 3 \times 10^{-6} = 3$ ppm. The people survive, except for those with asthma, or a bad cold, or any other condition that would be affected by almost lethal levels of SO_2.

However, in addition to the overall toxicity, we are specifically asked to find the concentration of SO_3; this is a gas phase problem, so the concentration becomes the pressure. For an estimate, we see that ΔG° is about −17 kJ, and as in the example earlier, this immediately gives us a large value for K_p. Most of the SO_2 is converted to SO_3. The estimated pressure of SO_3 is therefore around 3×10^{-6} atm. No matter what we do in the actual calculation, if we are very far from this, something is wrong. This is a case where doing the estimate is very valuable, as it is possible to make major mistakes if you just do the calculation without thinking about it. *Answer*: To one SF, this is the answer. However, we are asked to find the final concentration of SO_3, so we should do the calculation.

$$K_p = p(SO_3)/p(SO_2)p(O_2)^{1/2} = \exp(-\Delta G^{\circ}/RT)$$
$$= \exp(-(-16.7/2.5)) = 800.$$

The pressure of O_2 is 0.21 atm. so $p(SO)_3/p(SO_2) = 370$. The $p(SO_3)$ is therefore almost the total pressure of sulfur oxides, and if we have 3×10^{-6} pressure, we can check the number of moles to make sure:
$n = pV/RT = 3 \times 10^{-6} \times 5 \times 10^{13}L/25\ L - atm\ mol^{-1} = 6 \times 10^{6}$ moles. In all the rounding off, we lost a factor of 2, having started with only 3×10^{6} moles. As an exercise, try to find the missing factor of 2 — was it just rounding error?

COMMENT: The numbers in this problem were cooked to make the result interesting, so that we had to go through the calculation to know how it turned out. Suppose the inversion volume had been only 10 km^3? Then it would be a mass casualty event. Something like this happened in the 1950s in Donora, PA, which was a heavily industrialized town, with steel mills and other industry that produced SO_2. Note that 50 km^3 is pretty big — if the altitude of the inversion is 1 km, then, if the area is roughly circular, it would have a radius of close to 4 km, or a diameter of

8 km, or 5 miles. This is a pretty big inversion. There are many extra questions you could ask about this problem, which unfortunately is not entirely unrealistic.

4) Part 1:

$$O$$
$$\|$$

An ester (a compound of the form $R - C\text{-}O\text{-}R'$); hydrolyzes (reacts with water, often in acidic solution) to form a carboxylic acid and an alcohol; the reaction takes place in aqueous solution (R is a generic organic group):

$$
\begin{array}{ccc}
O & & O \\
\| & & \| \\
R - C\text{-}O - R' + H_2O & \rightarrow & R - C\text{-}O\text{-}H + R'\text{-}OH
\end{array}
$$

Call the original ester E, initial concentration 0.1 M, and call the products Ac (acid) and Al (alcohol). The concentrations of Ac and Al are necessarily equal; call this concentration C. Assume water concentration is included in K_C, and is constant (55.5 M >> 0.1 M, which is the most that can react, and we don't have enough SF to go further.) The equilibrium constant $K_c =10.0$ when water is included in the constant, so that it does not appear in the expression. The initial concentration, $C = 0$. Find the final concentrations.

Estimate: Since $K_c > 1$, we expect C to be less than 0.1 M, but not much less; obviously, it can't be more. The more difficult estimate is the concentration of E. However, let us assume that it is somewhere < 10% of the starting concentration, since K_C is large; the final concentration of E should therefore be < 0.01M; with K_c >>1, a value as small as 0.001 M cannot be ruled out.

Answer: We can write the equilibrium constant expression:

$$K_c = C^2/E$$

The concentration $[E] = 0.1 - [C]$. This makes the expression for $K_c = 10.0$

$$K_C = C^2/(0.1 - [C]),$$

This is a quadratic equation, which we can rewrite in standard form.

$$C^2 - 10(0.1 - [C]) = 0$$

Solve just using the standard solution of a quadratic equation, but be careful: this involves the small difference of two large numbers, and it will therefore be inaccurate if the square root in the solution is not carried to an extra figure. Then $C = 0.099$ M, and $E = 0.001$ M. Checking these values by plugging into the original expression for K_C, we get $K_C = 9.8 \approx 10$, which to one figure is correct. $C = 0.0991$ is too large.

COMMENT: We carried this solution to just one SF, in effect. What would usually just be round-off error is here exacerbated by taking a small difference of two large numbers. As a result, we lost considerable precision. The answer for C is between 0.0990 and 0.0991, and it is not worth going further.

Part 2: Suppose the value of Kc were 0.01 instead of 10. Repeat the problem

Estimate: Now we have mostly unreacted ester, so its final concentration should be larger than that of the product. Since we have one easy equation, $[C] + [E] = 0.1$, we could almost save time by not solving the equation at all, but simply starting with a plausible guess. Try $[C] = 0.03$

Answer: We have the same expression, but the value of K_C is now 0.01, and we get $C \approx 0.027$ M, and remaining reactant 0.073 M, to two SF; in this case, we can get a second SF. If we started with 0.03, and just tried to find the consequent value K_c; for C = 0.03, we would get $0.03^2/0.07 = 0.013$, so we just have a small adjustment, giving the 0.073 + 0.027 M result.

PART III: Next, suppose the reaction takes place in a solvent other than water, in which we can still define molarity. In this case take initial $[H_2O]$ to be 0.10 M and E = 0.10 M to start; we have to include water as a reactant in this case. For this solvent, $K_c = 0.1$

Estimate: $K_c = 0.1 = [C]^2/(0.1 - [C])^2$. This makes [C] / $(0.1 - [C]) = \sqrt{0.1}$, so $[C] \approx 0.03$ may be a reasonable first guess. ($\sqrt{0.1} = 0.316$). We need [C] to be close to 1/3 of 0.1 − [C]

Answer: [C] = 0.0241 M, [E] = 0.0759 M. The calculation, which is a simple linear equation, is left as an exercise. However, we have only two SF, so [C} = 0.24 M. [E] = 0.76 M

11

Thermochemistry: Basic Ideas of Thermodynamics; The Background of the Calculations on Equilibrium

There is a vocabulary that goes with the energy exchanges that happen in chemistry, as well as the rest of the universe. **Thermodynamics**, *the science of such energy exchanges, requires the introduction of a number of new variables. Here we introduce the definitions as well as a little of the use of these terms; we have already used one of these terms, free energy. However, this is intended to only be a brief summary, not a textbook, and it is expected that you will have a textbook for which this chapter serves to assist in solving problems using these concepts, especially by using approximations. In this chapter we see how these quantities relate to each other, especially in relation to chemical reactions. We finally get a discussion of free energy.*

In the course of anything that happens, there is an exchange of energy, in various forms of energy. We have introduced the use of *free energy* in earlier chapters, and promised a discussion in a later chapter. Now it is time.

System and surroundings: We are interested in what happens to a *system*, which can exchange matter or energy with its *surroundings*. If the system can exchange neither matter nor energy, it is an *isolated system.* If it exchanges energy with its surroundings, but not matter, it is a *closed system.* If it can exchange energy and matter, it is an *open system.* When we come to chemical reactions, we will mostly have closed systems. If you combine the system plus surroundings, we have the "universe" — everything we can be concerned with in the problem. The surroundings are generally considered large enough that when energy is added or taken from it, its condition is unchanged, so if the system gives heat to the surroundings, the changes in surroundings are *reversible*; we will define *reversible* as soon as we have distinguished path dependent properties from path independent properties.

Path dependent properties: Before we start, we should be clear about the difference between the *intensive property*, temperature and the *extensive property*, heat. A burning match has a much higher temperature than the ocean, but much less energy. The intensive property has much more energy in a local volume, but the total energy of the ocean is vastly greater because of its vastly greater mass. An intensive property is independent of the mass of the system, while an extensive property is proportional to the mass of the system. Next, we need to define the difference between path dependent and path independent properties. The heat is a path dependent property; it refers to a change in a system, and the heat is defined along the path between two states. For example, there would be a different heat if the system temperature changes suddenly and then continues at constant temperature, compared to a path defined by a gradual absorption of heat. The quantity of heat depends on how the heat is added. The heat is written as **q**. If heat is added to the system, raising its energy, it is counted as positive; when the system gives off heat, $\mathbf{q} < 0$.

There is also a second path dependent quantity. Much earlier, in the chapter on gases, we observed that pressure × volume had the dimensions of energy. In fact, you can change the energy of a gas by compressing it, or letting it expand. The energy $P\Delta V$ is the *work* (**w**) in this process. If the work is done on the gas (the system), it counts as positive — it adds to the energy of the system. The sum of heat and work is energy; energy is not path dependent; it is a *state function*, which means that it only depends on the state, not on how the state was reached. In other words, the change in energy is $\Delta E = \mathbf{q} + \mathbf{w}$, and is a state function. Heat and work, q and w, are path dependent, but their sum is not.

Equilibrium: If a system reaches a state in which it cannot change any more without some external input, it is at equilibrium. At this point, the *free energy* is a minimum; we have already used this idea in discussing osmotic pressure, and in discussing chemical equilibrium.

Thermodynamic state quantities: Those quantities that are not path dependent, like *enthalpy*, *entropy*, and free energy, as well as energy; these are quantities that depend only on the state. Now we have to define enthalpy and entropy, two new quantities. The only non-state quantities that we are concerned with are **q** and **w**.

Enthalpy and *Entropy*: These are thermodynamic state functions, like energy. Enthalpy is an energy term (it has the units of energy), and temperature times entropy is an energy term. These are quantities that are defined for any system at equilibrium (there is a way to define these quantities out of equilibrium, but we won't use that here). We can return to the path dependent quantities, **q** and **w**. If the transition in which the **q** and **w** occur is *reversible*, that defines a path; therefore $\mathbf{q}_{rev}$ and $\mathbf{w}_{rev}$ are defined on this path. This means the system moves along a path that is essentially always at equilibrium, which requires that it move infinitely slowly. A path that moves at a finite rate will have a different value of **q** and **w**. Enthalpy is more useful than energy in describing processes that occur at constant temperature and pressure, the most common conditions for chemical reactions, or at least for problems done using these concepts. In Chapter 7, we discussed heat capacity as one of the problems that could not be resolved in classical physics. Here we are not concerned with calculating heat capacity; we just assume that it exists. At constant pressure, adding heat to a system adds to the enthalpy, and the quantity of enthalpy change is the heat capacity x change in temperature, as discussed below; entropy will also be described below.

*The **first and second Laws of Thermodynamics:*** There are two Laws of Thermodynamics that are fundamental stability conditions for a universe. The **First Law** is simply stated: Conservation of Energy. There is no chemical process in which energy can be created or destroyed. (Ignore conversion of energy to or from matter, as the conditions for this are never achieved in chemical processes, but if we were concerned with the interior of stars, keeping track of energy would only require a bookkeeping adjustment.) This is a stability condition for a universe in that it forbids anything from popping in or out of existence. Whatever energy there is, is there forever, although its form may change. The **Second Law** is a little more complex: There are two statements of the Law, and the relation between them is not obvious, but it is not necessary to worry about this for the purpose of understanding chemical reactions. One statement, due to Kelvin, is that no machine can work in a cyclic manner without giving back some of the energy it takes in; this is necessary in order to return to the starting state so that it can take in more energy to go through another cycle. For the purpose of understanding chemical reactions, another statement, due to Clausius, is more relevant: In an isolated system, entropy reaches a maximum at equilibrium (Clausius's statement is actually closer to "entropy tends toward a maximum", and it really has to be defined for an isolated system, as adding energy, allowed in a closed system, would change the entropy). In a sense this can be understood as being equivalent to making the equilibrium state the most probable state. This too is a stability condition for a universe. For example, the Second Law drives mixing. If you have a room or other volume partitioned with gas on one side, vacuum on the other, and remove the partition, the gas will occupy the complete volume. If the volume does not have a partition, the gas will not return to one side. Such a return would not violate the First Law, but would have a probability so small that it would require a time much, much longer than the age of the universe. Obviously, a universe in which you could not be certain that the air around you would still be there in a second would lead to a very unstable existence. This never happens, and the Second Law is the reason. In this sense the Second Law, like the First Law, is a condition for a stable universe. Chemical reactions, and all other natural processes, also are governed by the requirement that entropy increases to a maximum.

The most relevant quantities for chemical reactions. In a brief introduction like this we can't include all of thermodynamics, so we will stick to energy quantities relevant to chemical reactions. We are primarily concerned with entropy, enthalpy, and free energy. However, before we get to chemical reactions, we need to list a few relations among these quantities, and especially to see how the quantities are measured. As always, without measurements, we cannot say anything. We start with measurements of certain quantities that are behind the entropy, enthalpy, and free energy that we will need to understand reactions.

Entropy: The heat is not a state function. However, if we choose a reversible path, then we can create a state function, with an "integrating factor" of $1/T$ (it is beyond the level of this course to define "Integrating factor" — just take it to be a multiplicative factor that makes a non-state function into a new function; in this case, it creates a state function). Thus $\Delta S° = q_{rev}/T$. Without defining a path, $\mathbf{q}$ is not defined. Because we use q_{rev} we have specified a particular value of $\mathbf{q}$ that exists along an equilibrium path; we can use it, with the integrating factor, to define a state function. A transition between states along a non-equilibrium path will have a value of q that cannot be used to define a state function. However, we can always find the $\Delta S°$ for a reaction, as a reversible path exists. While the reaction does not follow this path, it is defined and can be used to define $\Delta S°$. Since $\Delta S°$ is a state function it can be defined if a state is defined.

Heat capacity: One thing that can be measured is the amount of energy that is put into a known mass of a given substance, and the corresponding rise in temperature. The temperature rise for a given energy input depends on the substance. The amount of heat required to raise the temperature by 1 °C for a mole of the substance is the *molar heat capacity* (more chemical reactions than not take place at constant pressure, for which we use subscript p) $C_p°$:

For a given substance, $C_p°\Delta T$ gives the part of the enthalpy change that accompanies temperature change. Recall, in the chemical equilibrium discussion, how we introduced the free energy of formation. We can similarly define the enthalpy of formation $\Delta H°$ and an entropy of formation $\Delta S°$. There is an enthalpy of formation for each compound, $\Delta H°$, which is the enthalpy in the reaction forming the compound from its elements in their *standard states*. Usually, we take the standard state to be 298 K at 1 bar pressure see Chapter 10 for definition. While enthalpy is relative to a standard reaction, the elements have a standard *absolute entropy*, since there is a "Third Law of Thermodynamics" that sets $S° = 0$ at $T = 0$ K. The reason this is possible for entropy requires a discussion of the states that the system can access — more states means more entropy. In the chapter on spectra, we noted that more states were accessible at higher temperature; at absolute zero, only the ground state is accessible. With all the atoms or molecules in one state, we can set the entropy to zero, and we do. Once the entropy and enthalpy change has been defined, we can define the ΔG° of a reaction:

$$\Delta\Delta G^\circ = \Delta\Delta H^\circ - T\Delta\Delta S^\circ$$

where the first Δ means the change that happens during a reaction by summing over all the enthalpies and entropies of formation (the second Δ), reversing the sign of the reactants, which are used up. The reaction of course may not happen at standard temperature and pressure. We can find the difference in the enthalpy and entropy from the heat capacity if the temperature has changed, for example at constant pressure. If the pressure has also changed, there will also be PV work that must be included. It is possible to make these corrections from measured quantities. As all these are state functions, we do not need to be concerned about the path.

If we look at ΔH° we note the following: this is the heat transferred from the system to the surroundings. For the surroundings, the heat is $-\Delta H^\circ$ (the enthalpy change in the system is the negative of the heat change in the surroundings — what comes out of the system is added to the surroundings). This enthalpy is actually the *reversible* heat for the surroundings, since the surroundings are so large that the heat does not change its temperature or pressure. The reversible heat, q_{rev}, times the reciprocal of the temperature, is ΔS, so ΔH(reaction) is actually $-T\Delta S_{surr}$ (*i.e.*, $\Delta S_{surr} = -q_{rev}/T = -\Delta H$(reaction)$/T$) Then $\Delta G^\circ = \Delta H^\circ - T\Delta S^\circ = -T\Delta S_{surr} - T\Delta S_{sys} = -T\Delta S$ ("**universe**"), where universe means system plus surroundings. T is always positive, so maximizing the ΔS("**universe**") is the same as minimizing ΔG. For a chemical reaction, this means that the Second Law tells us that equilibrium is achieved by minimizing ΔG. Furthermore, the equilibrium constant is then

$$K_{eq} = \exp(-\Delta\Delta G^\circ/RT)$$

We do not derive this relation here — that has to wait for a more advanced course, but we will just say that this leads to the most probable distribution of reactants and products; we worked out some of the consequences of this in the previous chapter. This is a critical point to understand; it turns out that the driving force for essentially all changes at a molecular level is a drive to minimize free energy, or maximize entropy. This does not say that we can determine a rate from this, only an equilibrium. We must insert this caution here, because sometimes it is the relative rate of competing reactions that determines what actually happens; we talk about that in the next chapter. There are a couple of additional terms that need to be defined.

Endothermic, Exothermic: If $\Delta H > 0$, the reaction is *endothermic*; the reaction absorbs heat, which would make the local temperature go down; if $\Delta H < 0$ the reaction is *exothermic* and gives off heat.

For each thermodynamic quantity X, $\Delta X = \Sigma_j \, c_j \Delta_j X_j^\circ$ (products) $- \Sigma_k \, c_k \Delta X_k^\circ$ (reactants);
c_j is the stoichiometric coefficient of reactant or product j. When X is the enthalpy, this is called Hess's Law. X can be any state function, but not $\mathbf{q}$ or $\mathbf{w}$, which are path dependent. The thermodynamics of phase changes are a case of particular importance. In Chapter 9 we said that free energy at a phase change meant that the two phases had equal free energy. At a phase change, the switch from one phase to the other is essentially reversible. The key is that the free energy change across the phase boundary is zero. For example, at the melting point of ice, G(ice) = G(liquid water). Using the standard enthalpy and entropy values, at the phase change point,

$$\mathbf{\Delta S = \Delta H / T}$$

where the $\mathbf{\Delta}$ is the difference of the quantity between the two phases.

There is an interesting observation, known as Trouton's Rule, that for a liquid to gas transition, $\Delta S \approx 85 - 88$ J K^{-1} mol^{-1}. This holds for most substances, if they are not partially structured, as water is, for example. Why should this be? Entropy is a measure of the number of ways a system can be arranged (this is sometimes called a measure of disorder), and there are many more arrangements possible for a gas than for a liquid — thus the entropy of a vapor is much greater than that of the corresponding liquid. However, if the vapor is more or less like an ideal gas, the number of arrangements is independent of which gas it is — all ideal gases are essentially the same from this point of view. If the arrangements of the liquid are neglected, then we can understand why the different vapor phases, which have essentially the same entropy, give about the same phase change ΔS for so many cases. The exceptions are cases in which the liquid has some structure that is more important than in organic liquids, the most important case being water, which has $\Delta S = 109$ J K^{-1}mol^{-1}. In water the liquid phase has structure that is destroyed in the transition to the vapor phase, while the vapor phase is still more or less ideal, so there is a bigger difference. No comparable rule exists for solid $\rightarrow$ liquid transitions, as the number of arrangements of condensed phases can be more complicated, and varied. If the gas is monatomic, there are not as many states in the gas phase, and there is a difference in the other direction; liquid to vapor transitions in monatomic substances (e.g., neon, argon) have $\Delta S \approx 75$ J mol^{-1}K^{-1}.

Bond energies: We can associate an energy with a bond, and it is sometimes useful to consider bond energies. In a chemical reaction, bonds are broken and new bonds are formed, so one can get a rough idea of the reaction energy terms from the bonds that are broken and formed. Care is required in defining bond energies. However, the actual bond energies are easiest to define for a compound like CH_4 where all bonds are the same, and one can take the heat of formation, in this case divide by four, and attribute the result to the energy of one C–H bond. Here we will note that it is possible to make such a definition, at least approximately, but we will not use it.

With these relations and definitions, we can begin to do the problems that involve chemical reactions with their thermodynamic properties.

Key definitions: This chapter has included a set of definitions, and some relations among them. Since all the definitions and relations are in the text immediately above, we don't have to repeat them in the summary.

1) state and path
2) system, surroundings
3) equilibrium
4) heat and work; path dependent functions vs state functions
5) heat capacity; measurements
6) enthalpy, entropy, free energy
7) First and Second Laws
8) equilibrium constant, and its relation to free energy
9) endothermic, exothermic reaction
10) the consequences of the equality of free energy at a phase change

These definitions are tied to physical quantities that can be measured, like temperature and heat capacity. By understanding the relations among these quantities, we can at least begin to understand the physical processes driving chemical change. It will turn out that understanding the reason for chemical processes requires an understanding of the thermodynamics of the process. Obviously, we have not included all that would be in a full semester course in thermodynamics, nor have we used calculus, which makes many of the relations much easier to work with. What we do have makes it possible to see, at a basic level, what drives chemical change.

Let us see how this applies in calculations on reactions. If we have a general reaction, Reactants $\rightarrow$ Products, there will be an accompanying change in the entropy of "the universe", which we can write as a change in free energy. The change in free energy is $-T \times$ (change in entropy of the "universe"). With the minus sign, maximum entropy means minimum free energy. Let's call reactants Re, products Pr, the free energy difference ΔG, and at equilibrium, $\Delta G°$. Then when you get to equilibrium, we return to the relation we introduced earlier,

$$K_{eq} = [Pr]/[Re] = \exp(-\Delta\Delta G°/RT)$$

R, our universal gas constant (don't confuse this with reactants), is now best used in energy units that are appropriate for general use, not specialized for gases, so $R = 8.31$ J K^{-1}mole^{-1}. (If you conclude that 0.0821 L-atm/K mole corresponds to 8.31 J K^{-1}mole$_{-1}$, you are right; 1 L-atm K$^{-1} \approx 101$ J K^{-1}).

We can rewrite this equilibrium result as

$$-RT \ln (K_{eq}) = -RT \ln [Pr]/[Re] = \Delta G°$$

At equilibrium, there is no driving force on the reaction, so the difference in free energy of the pure substances, Pr and Re, is just balanced by the ratio of the concentrations at equilibrium, and this defines equilibrium.

To get some general sense of how this helps us in making estimates, we start with comparing RT with reasonable $\Delta G°$ values. At 300 K, $RT \approx 2.5$ kJ mol^{-1}. Chemical bond energies can be much larger than that, and the difference in energy in the reactants and products can easily be much larger. If $\Delta\Delta G°$ is much larger than RT and negative, then there is almost all product; if this quantity is much larger than RT and positive, there is almost all reactant; the reaction hardly begins. Only if there is a near equality in bond strength, so that $\Delta G°$ is of the order of RT, in magnitude, do we have comparable quantities of reactant and product.

The change in $\Delta G°$ with temparture makes it possible to determine $\Delta H°$ and $\Delta S°$ separately. For example, the slope of the $\Delta G°$ vs T gives $\Delta H°$. However, it requires calculus to gain anything useful from this, so we only note that this is the way to get these quantities, but we do not do any examples.

Summary: In the previous chapter we just gave the values of the equilibrium constant, and calculated the consequent concentrations of reactants and products. Let us now see what happens if we consider the enthalpies and entropies, as well as the free energy, as part of the problem, as well as related problems. This chapter has been very limited in its description of thermodynamics, only applying some relations to chemical reactions. Thermodynamics is a much broader subject, governing, for example, machines, and many other things.

Problems:

1)   The molar heat of formation of HCl is -92.31 kJ mol^{-1}. Find the heat produced when 50.0 g of H_2 are used with excess Cl_2 to produce HCl.

   Estimate: The equation for formation of HCl from the elements is:

   $H_2 + Cl_2 \rightarrow 2$ HCl, so there are 2 moles of HCl formed for each mole of H_2. There are almost 100 kJ produced per mole; 50 g of H_2 is almost 25 moles of H_2, so with 2 moles HCl per mole H_2, one gets 50 moles of HCl. Therefore, there should be less than $50 \times 100 = 5000$ kJ for 50 gm of H_2

   Answer: Here the problem has been worked out in the estimate — we only need to use the correct values, -92.31 kJ $\times 2 \times 50.0/2.016 = -4580$ kJ.

COMMENT: This seems to be almost too trivial a case for an estimate. All that is really required is more like a stoichiometry problem, in which the only thing that needs to be determined is the number of moles involved. That said, you still want to be sure that the energy is large compared with that for a single mole of H_2, and not orders of magnitude off if you somehow divide where you should multiply, or some such error. In the answer, we do have to be careful of SF, so we must use the atomic weight of hydrogen $= 1.008$, not just 1, if we are to get three SF.

2) 3.000 g of water vapor at 100.0°C condenses onto 100.0 g of liquid water, initially at 25.00°C.
Given: Heat capacity: C_p (liquid water) $= 4.184$ J g^{-1}. Enthalpy of vaporization of water at 100°C : $\Delta H_{vap} = 40.65$ kJ mol^{-1}

Part 1: Find the final temperature

Estimate: We are adding the heat of vaporization (as condensation, in this case, as the vapor is being converted to liquid, so the heat is given off, and is used to heat the 100 g of liquid (plus the 3g of vapor that is converted to liquid, but that is only about 3% of the total, so we ignore it for the estimate, although we will need to account for it when we get to the solution to four SF). Since we have the ΔH_{vap} per mole, we have about 3/18 as much per 3 grams. In this case, we have a choice: to do the problem per gram, or treat 100 g as a little more than 5 moles, multiply the per gram C_p accordingly, or do the problem in moles. Here, let's use grams—you can redo it with moles as an exercise. Then we have about 7 kJ to heat the water, and for 100 g, it takes a little more than 400 J $= 0.4$ kJ per degree to heat the water, so the water is heated about 15 to 20°C. (But there is a catch—see below).

Answer: $\Delta T = \Delta H_{vap} = (40.65 \times 1000 /18.016$J J g$^{-1} \times (3.000$ g$/103.0 \times 4.184)$ J °C^{-1} $(103.0$ g$)^{-1} = 15.71$ °C to 4 SF; The final temperature is therefore 42.71 °C. (BUT SEE COMMENT! This is the wrong answer.)

COMMENT: First, did we really need to know the number of grams of vapor to four places? Not in the final division by 103, but we need to know this for the amount of heat used to heat the water. But: are we really entitled to 4 SF? In fact, is this Answer even correct? No! Notice that the three grams of condensate start at 100 °C, and have to cool down, not be heated from 25°C up.

Answer: Equate the heat terms, as follows:
$(40650 \times 3.000/18.016)$ J $+ (3.000g \times (100 \text{ °C} - 25\text{°C} - \Delta T_f) \times 4.184$ J g^{-1} °C$^{-1}) = (\Delta T_f) \times 4.184$ J g$^{-1} \times 100.0$ g
Checking the units on each side, we see that each term has units of Joules—g and °C cancel wherever they appear. This check is necessary—it is too easy to make an error at this stage. The terms on the left are the J available from condensation, and the heat that comes from cooling the 3 g of water that starts at 100 °C. On the right is the heat used to warm the water starting from 25°C. Solving for the change in temperature, we get ΔT_f = 17.91 °C, for a final temperature of **42.91 °C**. Comparing with the erroneous earlier answer, we see a difference of only 0.2°C, which shows up in the third place. It is worth having the initial information to four places, when a conceptual error could allow the "correct" answer to two SF.

Part 2: as an exercise, do the same type of problem with melting ice; start with 100 g of water at 25°C, melt 10.00 g of ice at 0⁻°C into this water, and see what the final temperature is. Obviously, the procedure is the same, but the ΔH_{fusion} for the ice → water phase change is only 6.009 kJ mol^{-1}.

Estimate: Here the phase change ΔH is about seven times smaller, but the amount of ice that is given is more than three times as much as the vapor that was condensed in Part 1. It would be reasonable to guess that the final change in temperature should be roughly half as large, so the final temperature might be in the 16 to 17°C range.

Answer: This is left as an exercise. Follow the procedure used in Part 1, but be careful about signs; here the temperature has to go down.

COMMENT: This entire problem assumes that C_p is constant over the entire temperature range from 0°C to 100°C. This is not correct. If you know how to do an integral, look up the correct values, fit them to a function, and solve more accurately with the correct heat capacities. How does this change the relation between Part 1 and Part 2 (this is a question you should be able to answer even without doing the actual calculation.)

3)   Find the free energy of the reaction

$$2\ Al_2(SO_4)_3 + 12\ HCl_{(g)} \rightarrow 4\ AlCl_3 + 6\ H_2SO_4$$

Given the free energies of formation, in Table 11.1 below:

Estimate There is not much that can be done other than the actual solution here. It may however be worth checking the first SF, to guard against a gross arithmetic error. Be careful of signs. All the heats of formation are negative. Are the relations written so as to assume absolute values? Are the signs even consistent? This is a problem for you to solve. The products are, very quickly, $> -600 \times 10 > -6000$ kJ. The reactants are

Table 11.1.

	ΔG_f° (kJ mol^{-1})
$Al_2(SO_4)_3$	-3097
HCl	-95.3
$AlCl_3$	-628.8
H_2SO_4	-690.0

$2 \times 3000 + 12 \times 100 \approx -7200$. The reactants have greater negative $\Delta\Delta G_f^\circ$ by ≈ 1000 kJ, so the overall $\Delta\Delta G^\circ$ is > 0 for the reaction, but less than 1000 kJ.

Answer:

$$\Delta\Delta G^\circ = 4 \times (-628.8) + 6 \times (-690.0) - 2 \times (-3097) - 12 \times (-95.3) = 682 \text{ kJ}$$

Two points are worth noting: We round off the final answer to the nearest whole number, but carry the tenths places in the calculation. It is worth carrying the decimal places, as these amount to more than three kJ in a couple of places. However, the value for $Al_2(SO_4)_3$ does not have tenths of kJ; therefore, we don't know what the actual tenths value should be, so we must round off the final answer.

COMMENT: The answer to this problem is not really small, although it is the relatively small difference between much larger numbers. If you compare to $k_B T$, 2.5 kJ at 300 K, and even at 600 K it would be much less than the $\Delta\Delta G^\circ$ for this reaction. The equilibrium constant would therefore be tiny. However, here much of the energy comes from the $Al_2(SO_4)_3$, a large molecule with many bonds. The tiny equilibrium constant would correspond to a small amount in terms of the concentration of any molecule, and the tiny magnitude of the equilibrium constant still means that there is almost no reaction, even if we go to a more elevated temperature.

4) For the reaction $CH_{4(g)} + HCl_{(g)} \rightarrow CH_3Cl_{(g)} + H_{2(g)}$ at 298 K and 1 bar pressure,

The ΔG_f° are given by Table 11-2

Part 1: Find the free energy change for the reaction.

Estimate: Since the carbon compounds have similar ΔG_f°, what is left is HCl, so ΔG_f° must be close to $+100$ kJ (HCl is a reactant, so its sign must be reversed).

Answer: $\Delta\Delta G^\circ = \Delta G_f^\circ (CH_3Cl) - \Delta G_f^\circ (HCl) - \Delta G_f^\circ (CCl_4) = +87.6$ kJ

Table 11-2.

ΔG_f° (kJ)	
$CH_{4(g)}$	-50.79
$HCl_{(g)}$	-95.27
$CH_3Cl_{(g)}$	-58.5

Again, we are forced to round at the tenths position, as the value for CH_3Cl is not given to hundredths. $H_{2(g)}$ is an element in its standard state, so it has $\Delta G_f^\circ = 0$. So far, the fact that this is entirely a gas phase reaction has not been used.

Part 2: Now consider the reaction itself, but with each reactant's initial partial pressure = 10 atm, T = 298 K. The initial partial pressures of the products are zero. Find the final partial pressures. Assume the gases are ideal.

Estimate: The first thing we notice is that $|\Delta\Delta G^\circ| \gg k_B T$, and $\Delta\Delta G^\circ > 0$. Therefore K_p is extremely small, and the reaction barely gets started. $K_p = \exp\,(-87.6/2.5) \approx 6 \times 10^{-16} = K_c$ because $\Delta n = 0$. With such a small K_p the initial concentrations or pressures of the reactants essentially do not change, and the final pressures of both reactants are still 10 bar. Then the pressures of the products are going to be somewhere around the square root of $K_p \times$ (product of reactant pressures) or somewhere near 10^{-7}.

Answer: For a problem like this, one SF is probably adequate, since the values are so many orders of magnitude apart that minor corrections would probably affect the second SF in any case. Using the value of K_p from the estimate, let the product pressures both equal P_x (same stoichiometric coefficient), and have $P_x^2 = 100\,K_p$, so $P_x \approx 2.4 \times 10^{-7}$ bar. We are leaving the second SF, with the understanding that it is less than certain. This confirms that the initial pressures of the reactants are unchanged.

COMMENT: We are assuming ideal gases at 10 bar, which is likely to be at least a little off; at higher pressures gases are more likely to interact. At the 10 atm pressure, although there is hardly any partial pressure of product, the molecules still interact with the reactant molecules. Also, the values of ΔG_f° from standard states are used. Can we get away with it? Would the corrections be the same for reactants and products, considering the disparity in initial partial pressures? If these are ideal gases, we could get away with it, as there are no interactions to consider. For real gases, it is not so clear.

Part 3: Suppose 2.0×10^{-3} g of CH_4 actually reacts, with the initial pressures of 10 atm. for each reactant. Find the volume of the reaction vessel. The pressure no longer is fixed at 1 atm.

Estimate: This is a somewhat odd way to state a problem, but it is possible to solve it. Essentially, since we can equate the number of moles that react with the number of moles of product formed (same stoichiometric coefficients) we know the number of moles of product. From K_1 we know the partial pressure, which is what is relevant. With p, n, T we can solve for V, since we assume an ideal gas. We have roughly 10^{-4} moles at a pressure of about 10^{-7} bar, or atm – to this accuracy, it makes no difference; we should have around 10^3 molar volumes at 298 K and the partial pressure of product, or somewhere between 10^4 and 10^5 L. Also, for each mole of one product, we get one mole of the other, but we found the partial pressure for each product, so $n = 2.0 \times 10^{-3}/16$.

Answer: Write down the ideal gas law:

$$V = nRT/p = (2.0 \times 10^{-3}/16) \times 0.082 \times 298/\ 2.4 \times 10^{-7} = 1.3 \times 10^4 \text{ L}$$

We have 2 SF.

COMMENT: This is a trickier problem than it first appears and requires careful reading.

CONTINUATION: We could extend this problem by starting from the part III result that gives the volume, and change the temperature. What would the pressure be if the temperature were 333 K? there is a minor correction in the temperature in the ideal gas law, but a much more significant correction in the calculation of the value of K_p; $\Delta\Delta G°/RT$ at 298 K was 35.0, at 333 K it is 31.6, so K_p increases by more than an order of magnitude. Finish the problem as an exercise.

5) An "ordinary" liquid melts and boils with the following thermodynamic quantities: $\Delta H_m = 12.5$ kJ mol^{-1}; $\Delta S_m = 37.5$ J K^{-1}mol^{-1}; $\Delta S_b = 85$ kJ K^{-1}mol^{-1}. $T_b = 393$ K. Molecular Mass $= 120$ g mol^{-1} Find: 1) T_m 2) ΔH_m per mole. Subscript b = boiling, m= melting.

> *Estimate*: Here we have to use the fact that at the phase transition, $\Delta G = 0$, so $\Delta H/T = \Delta S$.
> For the melting point we are given ΔH_m and ΔS_m and have to find T_m; for the boiling point, we have T_b and ΔS_b and have to find ΔH_b. For the melting point, $T_m = \Delta H_m/\Delta S_m \approx 330$ K, and for the boiling point, $\Delta H_b = T\Delta S_b \approx 35$ kJ mol^{-1}. This is an estimate only to the extent that the answers have been written down without the use of a calculator.
> *Answer*: $T_m = 333$ K; $\Delta H_b = 33.4$ kJ mol^{-1}.

COMMENT: This is a plausible "ordinary liquid", with melting and boiling points in about the range that many organic liquids have them. If we found a boiling point above maybe 500 K, we might expect the liquid to have some strong intermolecular attractions that would make it other than ordinary.

6) Given the $\Delta H_f°$ values in Table 11-3: All molecules are in the gas phase.
    For the reaction (all molecules in the gas phase)

$$CH_4 + 2\,O_2 \rightarrow 2\,H_2O + CO_2$$

Table 11-3.

Molecule	ΔH_f° (kJ mol^{-1})
CH_4	-74.85
CO_2	-393.5
H_2O	-241.8

Part 1: Find $\Delta\Delta H$(**reaction**):

Estimate: We know how to do this kind of problem by now. The reaction is the combustion of methane, the main component of natural gas. Since we know that this produces a lot of heat, we expect the $\Delta\Delta H$(**reaction**) to be large and negative. Just looking at Table 11-3, we see three molecules on the right, each with $\Delta H_f^\circ > 200$ kJ, while the reactants have only one molecule, with $\Delta H_f^\circ < 100$ kJ, so without doing any work we expect $|\Delta\Delta H(\textbf{reaction})| > 600$ kJ and negative. O_2 is the element in its standard state, so its $\Delta H_f^\circ = 0$.

Answer: Just do the arithmetic: ΔH (reaction) $= \Delta H_f^\circ(CO_2) + 2\Delta H_f^\circ(H_2O) - \Delta H_f^\circ(CH_4) = -802.2$ kJ

Part 2: Burn 2 g of CH_4 in 100 L air at 1 atm pressure starting at 0°C. The heat capacity of air, both N_2 and O_2 is 1.0 J g^{-1} K^{-1}. No heat is lost; find the final temperature. This is called the adiabatic flame temperature.

Estimate: It's straightforward to find how much heat we have available to raise the temperature of the air — 2 g is 1/8 of a mole of CH_4 so we have about 100 kJ to heat the air. 100 L is about 4 ½ moles of gas starting at 0°C. The average molecular weight of nitrogen and oxygen in air is about 29 g mol^{-1}, so it takes 29×4 ½ ≈ 130 J $= 0.13$ kJ to raise the temperature 1°C; this would raise the temperature around 100 kJ/0.13 kJ°C$^{-1} \approx 750$°C.

NOTE: This looks like a big number, but not a big surprise — there is a lot of heat, and it doesn't take much to heat air — that is, the heat capacities of both oxygen and nitrogen are small. We know that burning natural gas should produce a high temperature, so the estimate, and the answer make sense. If we got a temperature rise of only about 100°C, we would suspect that something was amiss. Also, if the result were thousands of degrees, we would know something was wrong.

Answer: We have already done the hard part. What remains is to make the numbers as precise as the data we are given allows. The numbers in the estimate were very close as it happens — instead of 0.13 kJ, it should be 0.129 kJ. The final temperature to two SF = 780°C. The heat capacity is given as 1.0 J g^{-1} K^{-1}, so if this is ±5% (1.0 ±0.05, the error limit for the second SF), the temperature is also ±5%, and this could lead to an error of ±40°C. All the remaining small corrections can be safely ignored, since we are not going to get an accuracy sufficient for these other corrections to matter. The possible corrections if we had high accuracy would include the fact that the $\Delta\mathbf{H_f}°$ of oxygen is not exactly zero at 0°C (273K), as this is not the standard state — 298K is; also the ideal gas assumption is never entirely correct. We could look for other small corrections, but none matter when compared to the uncertainty in the heat capacity as stated.

****

A brief overall summary: The problems at the end of this chapter cannot cover all the concepts; they can only show some manipulations with certain thermodynamic variables. However, the change in thermodynamic variables, specifically free energy, $\Delta\mathbf{G}$, drives chemical change, in accord especially with the Second Law. There is a great deal of detail to be mastered to learn how to use these driving forces to obtain results for chemical reactions. However, the underlying idea is that the "universe" tends toward what is more probable, and this drives chemical and physical changes (we don't know enough to be sure that the quotes around "universe" are not needed). Once we understand this, we can use the apparatus of calculations to determine the extent of change, based on measurements of those quantities we are able to measure: heat, heat capacity, temperature, volume, pressure, and from these it is possible to derive the quantities that are not easily measured directly, like ΔS. What actually happens in a chemical system, however, does not always depend on the equilibrium; sometimes there can be more than one possible set of products, and which ones form depend on the relative *rates*. In chapter 13, we consider the rates at which these changes occur.

12

Solubility

In this chapter, we continue consideration of solutions, specifically with regard to the limit on the amount of solute that can dissolve. Solubility is defined as the amount of a substance, the solute, that can be dissolved in a given solvent to produce a *saturated solution*; the solution is saturated if addition of any more of the solute does not produce more that is dissolved. Instead, the added solute precipitates out of solution, and the concentration of the solute does not change. Solutions may of course have less than the saturation concentration, and the units of concentration are, as in previous chapters, especially Chapter 9, moles per liter (molarity, M), moles per 1000 g of solvent (molality, m), or even grams per 100 grams of solvent, which is sometimes convenient, but, as it does not involve moles, of limited use in most chemical calculations. If excess solute is added, meaning that there is undissolved solute in contact with the saturated solution, there is an equilibrium between the dissolved and undissolved solute:

$$S(\text{solution}) \leftrightarrow S \ (\text{undissolved})$$

If the solute is a salt, say MX, it would dissociate in an aqueous solution:

$$MX_{(s)} \rightarrow M^{+}_{(aq)} + X^{-}_{(aq)}$$

where the subscript (s) means solid, and (aq) means that the ion is in aqueous solution. If there is undissolved salt in contact with a saturated solution, then this is an equilibrium. If the solution is less than saturated, all the salt is dissolved, and the dissociation is effectively complete, although the ions, being of opposite charge, do interact with each other. Solutions with more than one solute are possible and the solutes may react, forming a new compound, which may be much less soluble, and lead to an entirely different solution. For example,

$$NaCl + AgNO_3 \rightarrow Na^+ + NO_3^- + AgCl_{(s)}$$

Because AgCl is almost insoluble, it precipitates, leaving a $NaNO_3$ solution. Of course, as written, this assumes equimolar amounts of NaCl and $AgNO_3$. In the more general case, there would be one salt in excess, so there would be the unreacted ions from that salt also remaining in solution.

As with any equilibrium, it is possible to define an equilibrium constant. For the salt define K_{sp} where sp means solubility product:

$$K_{sp} = [M^+][X^-] \text{ (at saturation)}$$

For example, suppose we start with a 1 M $AgNO_3$ solution. Then NaCl is added such that the final Cl^- concentration is 0.001 M. The solubility product of AgCl is 1.6×10^{-10}, so the remaining concentration of Ag^+ is $K_{sp}/0.001 = 1.6 \times 10^{-7}$ M.

We do not include the solid in the equilibrium constant, as it is a pure phase, and we can take it to have concentration 1, or we could think of it as incorporated into K_{sp}. Pure substances have unit concentration.

Le Chatelier's Principle: In an equilibrium, if you disturb the equilibrium, for example by changing the temperature, the equilibrium will shift so as to oppose the effect of the perturbation. If the equilibrium of a saturated salt solution is disturbed by adding a salt with one ion in common with the salt already present, that ion will form additional precipitate with the counterion (the oppositely charged ion). This leaves a solution with the common ion balanced with two counterions. In a sense, once we have the idea of an equilibrium constant, this Principle follows — adding reactant creates more product to keep the equilibrium constant unchanged, and adding more product similarly creates more reagent. Le Chatelier's Principle essentially states this in the form of a sentence, rather than an equation.

The ions do not have to be monovalent; in that case the equations above become:

$$M_aX_b \rightarrow aM^{+b} + bX^{-a}$$

This makes the charges balance. If you have a salt of the form A_2B_3 for example, you would need three plus charges on A and two minus charges on B; that way there are equal numbers of total plus and minus charges (in this case, 6 of each).

$$\text{Then define } K_{sp} = [M^{+b}]^a[X^{-a}]^b$$

We can also define a reaction quotient that has the form of K_{sp}, but has a value out of equilibrium, $Q = [M^{+b}]^a[X^{-a}]^b$

If $Q < K_{sp}$ the reaction is incomplete, and has to shift further in the direction of products; in other words the solution is not yet saturated, and can accommodate more solute. If $Q > K_{sp}$ the solution is probably supersaturated, and should return in the direction of reactants by precipitating the excess. If $Q = K_{sp}$ the reaction is at equilibrium, which means just saturated. This could be generalized to other reactions; we did not do this in the chapter on equilibrium, but the idea is occasionally useful.

Hydration: Water is a special solvent, and we will spend more time with aqueous solutions than with solutions in any other solvent. The amount of salt that can be dissolved in water is limited by the fact that each ion in the salt associates with (is "hydrated" by) several water molecules. For common ions like Na^+ or K^+, this is often around six water molecules. Five moles of salt in water, each mole having two ions, means that you have ten moles of ions altogether. Then, with six water molecules per ion, ten moles of ions means 60 moles of water, more than you have in 1000 g of water. It is therefore not a surprise that the limit of solubility for these simple salts is in the vicinity of 5 m. (At around 25°C the solubility of NaCl is about 6 m, of KCl about 5 m). The solubility of solids increases with temperature. The solubility of gasses decreases with increasing temperature. The solubilities of substances depend on their interactions with the solvent molecules; some ions hold their partners in the solid phase tight, and won't let go to interact with water. The solubility of some barium salts comes pretty close to one molecule per ocean.

Temperature dependence: *Solids are more soluble at higher temperatures, gases less soluble at higher temperatures*. One might rationalize this by noting that there are more degrees of freedom in gas than liquid, and in liquid than solid. More degrees of freedom means higher entropy, and higher temperatures favor the higher entropy phase. Higher entropy means lower free energy. This is very much a qualitative argument, and should not be taken too far, but it is reasonable.

Ion interactions: When we discussed colligative properties in Chapter 9, we considered the total number of particles in solution. We included a brief warning that the *effective* concentrations were not equal to the amount of a substance in solution, but were modified by the interactions among the ions or other molecules, interactions among ions being almost always most important. This effect becomes apparent when we measure, say, freezing point depression, or osmotic pressure; we find that there seem to be fewer particles in solution than would be expected if there were full dissociation of salts. This can be accounted for by considering the interactions among the ions. The cations have a little more negative charge, in the form of negative ions, surrounding them, and the negative ions a little more positive charge. In a very dilute solution, it is possible to calculate this effect, and it was done in the 1920s by Peter Debye and his student Ernst Huckel. Not surprisingly, the energy of ion interaction appears in the calculation in an exponent, as $\exp(-E/RT)$. If $E \ll RT$, the exponent can be expanded and the equation solved. However, for more concentrated solutions the energy is too large. In this case, we have to jump ahead a century to more or less the present time, and use a large computer to simulate the system. However, it is possible to determine the magnitude of the effect experimentally by quantitatively measuring the decrease in osmotic pressure or the freezing point depression. This gives the effective concentration, known as the *activity* of the solute, and especially of the ions.

Solubility of ions: Different ions have different tendencies to stay in solution. For aqueous solutions, this depends in part on their ability to hydrate, compared with the energy required to break the ionic bonds holding them in the ionic solid. The alkali cations, Na^+, K^+, etc. tend to be soluble with essentially any anion. Many halogen anions are fairly soluble to very soluble. Nitrates, NO_3^-, are all soluble. Alkaline earth (Mg^{2+}, Ca^{2+}, etc.) metal ions are soluble to some extent depending on how well the ions can be hydrated. Silver chloride is largely insoluble. On the other hand, some ions are so tightly bound to each other that water has no hope of hydrating them. Bi_2S_3 supposedly dissociates, and is thus soluble, to the extent of something like one molecule per ocean. This may be a slight exaggeration, but then again, maybe not. There are other ionic compounds, with intermediate solubilities. Heavier ions are harder to hydrate (or have less energy of hydration) so are often less soluble. For example, the solubility of AgCl > AgBr > AgI; AgI has $K_{sp} \approx 10^{-16}$, compared to the roughly 10^{-10} for AgCl.

Summary: Ions dissociate in solution, and it is possible to work out the equilibria that accompany these solutions. There is a limit to the amount of any salt that can dissolve; in aqueous solution, this is set in part by hydration; if essentially all the water is occupied by hydration, then it is no longer possible for more to dissolve. For some salts, hydration is weak, but the bond between the ions is relatively strong, so very little dissolves.

Problems:

Problem 1) AgI has $K_{sp} = 1 \times 10^{-16}$. Find the number of grams of Ag^+ in 237 mL of a saturated AgI solution.

Estimate: If there is only one salt in solution, and it is a 1:1 salt, the concentration is $\sqrt{K_{sp}}$, here 10^{-8} M. We have about 0.2 L, and the atomic mass of Ag is about 100, so we expect around $0.2 \times 100 \times 10^{-8}$ order of magnitude of $= 10^{-7}$ g. In doing this estimate, if we got more than an order of magnitude away from 10^{-8} g we might suspect that we had an error. Weights and liters are only an order of magnitude away from 1. If the volume given were on the order of mL, we would expect less; there is no way the atomic mass could be >1000.

Answer: We already used the formula volume x atomic mass x molarity in the estimate. If you have a problem with figuring out the formula, check the units: $L \times g\ mol^{-1} \times mol\ L^{-1} =$ g; the moles and liters cancel. If all else fails, this is a good check. Answer $= 3 \times 10^{-7}$ g. Because the K_{sp} value is given to only 1 SF, this is the best we can do with the answer.

Problem 2) For PbI_2, $K_{sp} = 8.4 \times 10^{-9}$. Find the number of grams of Pb^{+2} in 237 mL of saturated PbI_2 solution.

Estimate: This differs from Problem 1 in two ways: K_{sp} is given to two SF, and, much more important, there are two I^- ions, so we have $K_{sp} = [Pb^{2+}][I^-]^2$ –There are 2 I^- ions for each Pb^{2+}. With three ions we will need a cube root, and the K_{sp} value is already orders of magnitude larger than in problem 1, so we know at once that the answer will be orders of magnitude larger than in problem 1. If $[Pb^{2+}] = x$, then $[I^-] = 2x$. and $K_{sp} = 4x^3$. This means that x should be of order 10^{-3} M $\approx (K_{sp})^{1/3}$

With atomic mass of 207 and 0.237 L we multiply x by about 50, so we expect roughly 0.05 g (200 g mol^{-1} $\times$ 0.240 L $\times$ 10^{-3} mol L^{-1}). With the much larger K$_{sp}$ and the cube root, we were expecting a much larger answer than in problem 1. What we said in problem 1 about checking units applies here as well. *Answer*: 207 g mol^{-1} $\times$ 0.237 L $\times$ 1.28 $\times$ 10^{-3} mol L^{-1} = 0.063 g. Here we can keep two SF and we have an answer that is in fact orders of magnitude larger than the answer we found in Problem 1.

Problem 3) A salt, ZX, has K$_{sp}$ = 10^{-10}. There is a reaction

$$AX + C^- \rightarrow AC + X^-$$

AX and AC are soluble as molecules and do not ionize so the soluble anions come from other substances. The reaction has an equilibrium constant of 5.0. Consider a solution initially 10^{-4} M in C$^-$, which contains 10^{-6} M of Z$^+$ as an impurity.

Part 1: How much AX must be added to produce a precipitate of ZX?

Estimate: The reaction must produce 10^{-4} M = 10^{-10}/10^{-6} of X$^-$ to get a precipitate. Since we have 10^{-4} M of C$^-$, the reaction would have to go to completion — that is, it would need to have an equilibrium constant of infinity — all the C$^-$ would have to react to produce 10^{-4} ZX. While the equilibrium constant is appreciably greater than 1, the reaction still does not go all the way to completion, so there is nothing to calculate; no amount of AX will produce a precipitate of ZX, because there is only 10^{-4} M C$^-$ available to react.

Answer: The estimate is the answer. So far, there is no reason to set up equations or to do any calculations.

Part 2: Suppose the solution is 10^{-3} M in C. Now how much AX reacts to get a ZX precipitate?

Estimate: This time the question can be taken seriously, as there is enough C$^-$ to produce a precipitate. Clearly you have to add more than 10^{-4} M AX in order to get the 10^{-4} X^- that is needed to produce the precipitate. Since the equilibrium constant for the reaction is > 1, we won't need orders of magnitude more than 10^{-3} M. Therefore, we have a lower bound, 10^{-4} M, and almost certainly an upper bound of 10^{-3} M. At the level that just produces a precipitate, $[X^-] = [AC] = 10^{-4}$. It remains to solve for required AX. The amount that we find will be the amount that is left after 10^{-4} has reacted, so the answer is 10^{-4} + something; the something can be smaller than 10^{-4}, so the answer may turn out to be just 10^{-4} to the number of SF that we have (2).

Answer: $K_{sp} = 5 = [X^-][AC]/[C^-][AX] = (10^{-4})^2/(0.9 \times 10^{-3})$ [AX]

where we have set $[C^-]$ at the value it would have if the necessary 10^{-4} M X^- is produced — this would use up 10^{-4} M C^- out of the original 10^{-3}M, leaving $0.9 \times 10^{-3} = [C^-]$. Now we can solve

$$[AX] = 10^{-8}/5.0 \times 0.9 \times 10^{-3}$$
$$= 2.2 \times 10^{-6}$$

This is the amount we add to the 10^{-4} that has reacted; it is the required amount that is left unreacted. As we estimated, to 2 SF, the answer is 1.0×10^{-4} M.

Problem 4) Radium (Ra) is radioactive, and the result of radioactive decay is a gas, Radon (Rn). 10 L of a *saturated solution* of $RaSO_4$ ($K_{sp} = 4 \times 10^{-11}$) is connected to an apparatus that collects the radon gas emitted by the radioactive decay of the radium. The half-life of radium is 1500 years (half-life is defined as the time for half of the isotope to decay), so we can reasonably ignore any changes in quantity of radium due to this decay in a month, but we need to know this to get the rate of production of Rn.) Find the volume of

Rn collected at 25°C and 1 atm, after 1 month. We can use the rate relation

$$\text{Moles(time} = t) = \text{moles(time} = 0)\exp(-0.693t/t_{1/2})$$

$t_{1/2}$ = half life, here 1500 years. $0.693 = \ln 2$, and is necessary in order to use half-life with the exponential – essentially it converts base 2 to base e (=2.718).

Estimate: We can estimate the rate of production of Rn, and assume it is constant over a time very short compared to its half-life. We know how much Ra we have, and $K_{sp} = 4 \times 10^{-11}$; as $RaSO_4$ is a 2:2 electrolyte (each ion has two charges, making the ratio of ions 1:1), the concentration of each ion is the square root of K_{sp} or about 6×10^{-6} M. In 10 L, there are therefore 6×10^{-5} moles. Half the amount of Ra would disappear in 1500 years = 18,000 months, so the amount of Ra that decays is about 1/36,000 of the amount that it starts with. We really don't need the equation for the exponential decay, because the time periods involved are so different — the exponent is too tiny to matter. Therefore we estimate roughly 6×10^{-5} /$4 \times 10^{4} \approx 1.5 \times 10^{-9}$ moles in one month. Since the molar volume under the conditions specified is about 25 L, we expect around 4×10^{-8} L = 0.04 μL.

Answer: We just have to get a two SF determination of the amount of Rn formed. The concentration of the saturated solution is, to 2 SF, 6.3×10^{-6} M, so the amount of Ra is 6.3×10^{-5} moles in 10 L. Because the collection time is <<1/2 life, we can linearize the exponent, and assume that the rate is constant. One month is 1/18000 of half the Ra, so 6.3×10^{-5} / 36000 = 1.7×10^{-9} moles of Rn is formed, The molar volume at 25°C and 1 atm is 24.6 L, so we have 1.7×10^{-9} moles $\times$ 24.6 L mol^{-1} = 4.3×10^{-8} L = 0.043 μL.

COMMENT: In this case it is hard to have any intuitive feel for a reasonable answer. The numbers are far away from our normal sense of what is to be expected. The calculation of the amount of Rn produced is correct — but, does it really make any sense? Not at all. Aside from the difficulty of collecting such a small amount of gas, there is a question of whether there is any gas at all. The solubility of Rn in water is about 230 mL (gas) L^{-1} (water), way more than is produced in the decay of the available Ra. In principle one could pump off the water, and the dissolved other gases, all of which are far less dilute than the Rn, but this would be extremely difficult. The problem on its face looks realistic, but it is actually not plausible.

SUMMARY COMMENTS: This is a small number of problems, but it is sufficient to illustrate at least some of the range of questions that can come up in solving solubility problems. The most important takeaway from these problems, just as, to a large extent, those in the other chapters, is that there may sometimes be order of magnitude differences in the quantities being calculated. If you are adding two concentrations that differ by three orders of magnitude, and you only have two SFs, throw away the smaller quantity. Of course, in multiplication/division, this does not apply. By taking advantage of these differences the problems can be simplified. Also, in checking the problem's solution, the orders of magnitude, and the units, at least should be consistent; by checking the units as you go, you can make sure that there is not some major glitch in what you are calculating. In addition to all this, you can ask whether the problem's conditions are physically reasonable. Some problems can be stated in a way that appears perfectly reasonable, but an examination of what must be happening shows that the conditions stated in the problem could not possibly exist. The problems actually require consistency of units, *and physically reasonable conditions. Then there may be simplifications that can be applicable and make solving the problem much easier. Sometimes there can be multiple equilibria in a single problem, in which one can either use some sort of software with a computer, or use the simplifications that are available. If all the equations are linear, having multiple unknowns is not so difficult, if you can diagonalize a matrix (not always trivial, if more than 3×3; also, diagonalizing a matrix may require more math than you have at this point). However, many of the equilibrium problems in this chapter are non-linear. Those in the next chapter, which are more general than those in this chapter, are even more likely to be non-linear. To repeat, look for simplifications based on the values of quantities, keep track of units and significant figures, and see whether the physical conditions are reasonable. Solving first to one SF is often useful, as all of these checks can be made to 1 SF (except for keeping track of the significant figures), and the danger of getting lost in the arithmetic is greatly diminished.*

13

Acids, Bases, and pH

Introduction: We are going to primarily discuss aqueous solutions. Water is a marvelous solvent. It is not an ionic molecule, although it is not entirely neutral across the molecule. One end is more nearly negative, near the oxygen atom, the other end has balancing positive charge near the hydrogen atoms. Water is an example of a *polar* molecule; it has a positive end (pole) and a negative end (pole). A very small fraction of the water molecules can actually ionize:

$$H_2O \rightarrow H^+ + OH^-$$

Actually, this reaction never happens; instead, the reaction is closer to

$$2\,H_2O \rightarrow H_3O^+ + OH^-$$

The hydrogen atom has only one electron, and if it is lost, what remains is just the nucleus, which is just a proton. The proton is orders of magnitude smaller than a full atom, and the electric field around a charge goes inversely with distance. In other words, the bare proton would have an immense field, and would immediately attach itself to another molecule, making H_3O^+. This is a molecule with normal size and exists in solution. However, the molecule can trade its extra proton with another water molecule, so we cannot label the individual molecules. Quantum mechanics tells us that identical particles are indistinguishable, so this is not even a real question. H_3O^+ is called the **hydronium** ion.

In what follows, we may sometimes write H^+ for convenience, as we do in the next paragraphs, but this always means H_3O^+. Later, we will note that the H_3O^+ itself tends to attach to either one more water (Zundel ion) or three more (Eigen ion).

There is an equilibrium constant for the ionization of water, and we write it as an ion product, since the concentration of water is not affected by the ionization to the extent that it is measurable, so we write instead, at 25 $^\circ$C, 1 atm.,

$$K_W = [H_3O^+][OH^-] = 1.016 \times 10^{-14},$$

giving the molar concentrations in pure water at 25 $^\circ$C, $[H_3O^+]=[OH^-] = 1.008 \times 10^{-7}$ M. In other words, this omits the $[H_2O]$ that the correct reaction equilibrium constant would have. The numerical consequences of omitting it are not important to any reasonable approximation (the value of the constant would be different, but as long as it remains constant, we can use the more convenient form); the K_W form is much more convenient. Also, we ignore the temperature dependence of K_W and just write it as 10^{-14}, with $[H_3O^+] = [OH^-] = 10^{-7}$. Furthermore, at this level we ignore differences between molar and molal. As pure water is 55.5 M, the fraction that ionizes is $10^{-7}/55.5 = 1.8 \times 10^{-9}$; about two molecules of water in one billion ionizes.

K_W is proportional to a proper thermodynamic equilibrium constant, so it behaves in the same way, and the product $[H_3O^+][OH^-]$ is constant; if the concentration of one of the ions increases, the other must decrease. The ion concentrations can range from >1 M to $< 10^{-14}$M, as long as the other ion changes to keep the product equal to 10^{-14}.

Definitions of acid and base: There are three definitions of acid and base that are generally used. Arrhenius defined an acid as a molecule that could give up H^+. This turns out to be somewhat limiting. A more practical definition is due to Bronsted (or sometimes called the Bronsted-Lowry definition): an acid is a proton donor, a base is a proton acceptor. There is also a Lewis acid, which defines an acid as an electron acceptor, while a base is an electron donor. This is the most general definition, as it allows compounds to be acids and bases that do not contain hydrogen or hydroxide, or could not donate or accept any that may be present, discussed as.

However, for aqueous solutions, the Bronsted definition is most useful. Then we can write an acid as HA, which ionizes in water as

$$HA + H_2O \rightarrow H_3O^+ + A^-$$

This allows us to define an acid constant, K_a as

$$K_a = [H^+][A^-]/[HA]$$

Here, we abbreviate H_3O^+ as H^+, and omit the constant value $[H_2O]$. Thus for a base that accepts a proton we have

$$B + H_2O \rightarrow BH^+ + OH^- \text{ (equivalent to } B + H^+ \rightarrow BH^+\text{)}$$

$$K_b = [BH^+][OH^-]/[B],$$

Multiply by $[H^+]/[H^+]$ to get $K_b = [BH^+][OH^-][H^+]/[B][H^+] = K_W/K_a$ where the K_a is that of the *conjugate acid* of the base B, that is, BH^+. We discuss conjugate acids or bases in the next paragraph, along with *hydrolysis*, the reaction with water. There is OH^- produced here, so the solution becomes more basic, or less acid.

Some acids and bases are strong, meaning that their constants are essentially infinite — they react completely, leaving no HA or B, but only A^- or BH^+. However, there are also weak acids and bases, and these do not react completely, leaving a finite amount of HA or B; in fact, unless the large majority of the molecules remain unionized, we do not count the acid or base as weak. The *conjugate base* of the weak acid HA is A^-, and the conjugate acid of the weak base B is BH^+.

A strong acid will transfer its proton to the weaker conjugate base, and a strong base will take a proton from a weak conjugate acid. We could write an equation comparable to the ones above for the weak conjugate acid or base, e.g.,

$$BH^+ \rightarrow B + H^+$$

$$K_a^* = [B][H^+]/[[BH^+]$$

$$\text{Then } K_a K_b = [BH^+][OH^-]/[B] \times [B][H^+]/[[BH^+]$$
$$= [H^+][OH^-] = K_W = 10^{-14}$$

where K_a^* is the acid constant for the conjugate acid of the weak base. In other words, this repeats the derivation above in a slightly different way, but leads to the same result. We could write the corresponding equations for the conjugate base of the weak acid.

The reaction of the conjugate base of the acid with water (this looks like what we saw in the preceding paragraph)

$$A^- + H_2O \rightarrow HA + OH^-$$

The equilibrium constant for this reaction is also called the *hydrolysis constant* K_h. If a salt, say sodium acetate, is added to water, the acetate anion will react with water according to that equation, as acetate is the anion of a weak acid, acetic acid. The equilibrium constant for that reaction is what we have just written (or rather, the acid equivalent), so

$$K_h = [HA][OH^-]/[A^-]$$

If we multiply this by $[H^+]/[H^+]$, we get $[HA]/[H^+][A^-] \times [H^+]$ $[OH^-] = K_W/K_a$

$$\text{So } K_h = K_W/K_a$$

The sodium acetate solution will be basic — the OH^- produced in the hydrolysis reaction makes $[OH^-] > [H^+]$. As always, $[H^+][OH^-] = 10^{-14}$, so the addition of hydroxide means the hydronium concentration must decrease accordingly. NaOH is a strong base, which ionizes completely. The salt of a strong base and a weak acid makes the solution basic because of hydrolysis. Similarly, the salt of a strong acid and weak base makes the solution acid. For example, NH_4Cl makes the solution acid (NH_4OH is a weak base), while HCl is a strong acid. Weak bases are typically amines (nitrogen compounds). NH_4OH is the product of the reaction of ammonia, NH_3, with water and apparently cannot be isolated as a separate compound out of aqueous solution.

At this point it is time to introduce pH, which is the usual way of discussing acidity of a solution.

$$pH = -\log_{10} [H^+], \text{ or we can write } [H^+] = 10^{-pH}$$

We can also define a $pOH = -\log_{10}[OH^-]$. This is not often done; you would have $pH + pOH = 14$, so pOH is essentially redundant.

BUFFERS: Buffer solutions have ions that can take up either H^+ or OH^- and thus prevent the pH of the solution from changing too much when more acid or base is added. Blood is an example — if the pH strays too far from around 7.3 to 7.4, the patient is in trouble.

We can produce a buffer from either a salt of a weak acid and its conjugate base, or the salt of a weak base and its conjugate acid. For example, consider a solution of an acid like acetic acid, HAc, and its conjugate base, acetate ion. We can prepare such a solution by mixing sodium acetate and acetic acid. Then if base is added to the solution, it will react with the HAc, producing some more acetate ion, but changing the solution pH very little. If acid is added, it will react with the Ac^- making the solution only very slightly more acid. Only if so much acid is added that it reacts with almost all the Ac^- or so much base that it reacts with almost all the HAc is there a major change in pH. The amount of acid or base that can be added without using up all the acid or base is the *buffer capacity*.

For example, suppose there is a solution with $[HA] = 0.300$ M and $[Ac^-] = 0.200$ M. The pK_a, defined as $-\log_{10} K_a$ of acetic acid is 4.8, or $K_a = 10^{-4.8} = 1.6 \times 10^{-5}$.

$$\text{Then } K_a = 1.6 \times 10^{-5} = [H^+]0.200/0.300$$

$$\text{So } [H^+] = 2.4 \times 10^{-5}, \text{ and pH} = 4.6$$

If the $[HA] = [A^-]$ the pH would be 4.8. We always consider the amount of H^+ and OH^- to be small enough to neglect in comparison to the concentration of the acid or base and salt. Suppose we originally had $[HA] = [A^-] = 0.25$ M, and then added 0.05 M of strong acid, making the concentrations those we just used. Without the buffer, 0.05 M strong acid gives $pH = -\log[0.05] = 1.3$, obviously a huge difference. Only if we added more than the 0.25 M amount of buffer, exceeding the buffer capacity, would we see such huge changes. However, if we consider the change in terms of the actual quantity of hydrogen ions, it does not look so small. $10^{0.2} = 1.58$, so a change of 0.2 pH units means a change of 58% in $[H^+]$.

Titrations: Suppose we had to determine the concentration of a solution of acid or base. We could *titrate* the solution. For example, suppose we had an unknown solution of HCl, and a known solution of NaOH, with concentration 0.1103 M (it is more trouble than it is worth to get a solution to be something convenient like 0.1000 M; as long as we know the concentration, it is easier to do the arithmetic than to prepare an exact value of a concentration in lab.) If the solution is even 10^{-4} M in acid, it has pH 4; add a fraction of a drop of concentrated base, and you might have 10^{-4} base, pH 10. This jump is easily detectible with an indicator (a dye, like litmus, that changes color at some intermediate pH) or a voltmeter, with an electrode that responds to change in pH, a topic we will get to in the chapter on electrochemistry. If you can measure the volume to half a drop = 0.02 mL, you can determine the concentration with adequate accuracy.

The fundamental idea is one we have met already, that volume times molarity = volume $\times$ molarity = amount (moles) present; the titration endpoint comes when the number of moles of acid equals the number of moles of base.

That is, acid moles $L^{-1} \times L$ = base moles $L^{-1} \times L$ = moles

When we do a titration the molarity of one of the substances, say the base, is known, as is the volume of the base, so we know how many moles of base we use to get to the endpoint of the titration. We know the volume of acid we used at the start, so the only unknown is the concentration of acid, which is easily solved for. We should be careful to use base that is sufficiently dilute that we get something reasonable for the volume. If the volume is too small, 0.02 mL won't be small enough compared with the total volume to give the required accuracy. Suppose the volume turns out to be 20.00 mL, so an error of 0.02 mL gives an error in the result of one part per thousand. If there were only 2.00 mL, the error would be ten parts per thousand, which is generally too large for traditional analytical chemistry, at undergraduate lab level.

Charge and mass conservation: There is another relation that must be satisfied with charged particles; charge must be balanced. The sum of the concentrations of positive ions x the charge on each must equal concentration of negative ions x the charge on negative ions — overall, the solution must be neutral:

$$\Sigma\, q_j^{+} c_j^{+} = \Sigma\, q_j^{-} c_j^{-}$$

where q_j^{+} is the charge on the jth positive ion, and c_j^{+} the corresponding concentration; the negative charged ions have the corresponding terms on the other side of the equation. For example, suppose we have a system consisting of water with added Na_2HPO_4 in which Na^{+} has one charge, and HPO_4^{-2} has two. However, if you start with HPO_4^{2-} it can hydrolyze (react with water: $HPO_4^{2-} + H_2O \rightarrow H_2PO_4^{-} + H^{+}$) to give $H_2PO_4^{-}$ as well, so this has to be included in the group of variables. Then we have an equation for charge

$$[Na^+] + [H^+] = [H_2PO_4^-] + 2[HPO_4^{2-}] + [OH^-].$$

A sample problem: Continuing with phosphoric acid ionizations, there are the equilibrium equations for the hydrolysis of $[H_2PO_4^-]$, $[HPO_4^-]$ that comes from the ionization of $[H_2PO_4^-]$, the $K_W = [H^+][OH^-]$ equation, and the conservation of mass. pK_a for $H_2PO_4^- \rightarrow H^+ + HPO_4^{2-} = 7.2$. We also know how much Na_2HPO_4 is added, and this produces the total concentrations of HPO_4^{2-} and $H_2PO_4^-$. This makes five equations for five unknowns: Na^+, HPO_4^{-2}, $H_2PO_4^-$, H^+, OH^-. We have already made one approximation here, ignoring PO_4^{3-}, assuming (based on our knowledge of the pK_a values) that by starting with $H_2PO_4^-$ we will get a negligible amount of PO_4^{3-}. If we had enough SF to include this ion, we would need a sixth equation, which would be the pK_a of HPO_4^{-2}; however, for any reasonable set of conditions, this assumption is valid (see the question below). These equations are not all linear. Obviously, approximations are needed. Suppose we start with enough Na_2HPO_4 that the phosphate ions and the sodium ions are ≈ 0.01 M, or greater; specifically, in this case, let us start with 10.00 g Na_2HPO_4 in 1.00 L water. Also, if the pH is at least 1.5 units from 7, the amount of one of the ions H^+ or OH^- will be negligible compared to the other. The five equations are:

1: $K_{a2} = [H^+][HPO_4^{-2}]/[H_2PO_4^-]$ (acid dissociation of $H_2PO_4^-$; $K_{a2} = 10^{-7.2}$)
2: $K_h = [H_2PO_4^-][OH^-]/[HPO_4^{2-}] = K_W/K_{a2}$ (hydrolysis of HPO_4^{2-})
3: $K_W = [H^+][OH^-]$
4: $[Na^+] + [H^+] = [H_2PO_4^-] + 2[HPO_4^{2-}] + [OH^-]$ (charge conservation)

5: $(NaH_2PO_4)_m / [mNaH_2PO4]_o = \{[HPO_4^{2-}] + [H_2PO_4^-]\} \times V$
(mass conservation; on the left we have initial moles of added (NaH_2PO_4), on the right the amount actually in solution, all in moles; for the left side explanation: see text below)

In equation 2, we multiplied the equilibrium expression on the right by $[H^+]/[H^+]$ to get K_w/ K_{a2}. In equation 5, $[mNa_2HPO_4]$ is the molecular mass of Na_2HPO_4 and $(Na_2HPO_4)_m$ is the mass of Na_2HPO_4 added initially; in other words, the moles of Na_2HPO_4 added to the solution to begin with. This must end up as one of the two hydrogen phosphates, The right hand side is the molarity, moles L^{-1}, times the volume of solution, V, in L. This is mass balance; the number of moles of phosphate you start with on the left equals the number of moles you end with on the right. We could complicate matters still more by observing that there are two more equilibria that could in principle play a role, each involving a new species. $H_2PO_4^-$ could hydrolyze, giving H_3PO_4, and HPO_4^- could ionize giving PO_4^{3-}. However, let us start by assuming that these make a negligible contribution; we can check this at the end. This is a problem in which an estimate is absolutely necessary, unless you have a computer handy. The equations are non-linear, so you cannot even solve by simply diagonalizing a matrix, assuming you know how to do this and have appropriate software. Start with conservation of charge, and assume there is one dominant equilibrium, the hydrolysis of HPO_4^- (we could equally well have written this as the ionization of $H_2PO_4^-$). If this gives a pH and a consistent set of concentrations, then we can use it to find the concentration of the minor constituents.

$$HPO_4^{-2} + H_2O \rightarrow H_2PO_4^- + OH^-$$

With two charges on the left and two on the right, this conserves charge. Its equilibrium constant is

$$K_h = [H_2PO_4^-][OH^-]/[HPO_4^{2-}] = K_w/K_{a2} = 0.13 \times 10^{-6}$$
(equation 2 above).

If we ignore everything else, $[H_2PO_4^-] = [OH^-]$. Let $[OH^-] = y$. Then $K_h = y^2/([HPO_4^{2-}]_0 - y)$ where $[HPO_4^{2-}]_0$ is the initial concentration of HPO_4^{2-}, which we know is 0.07004 M for 10.00 g L^{-1} of Na_2HPO_4. This reduces to a single quadratic equation. Solve, with $K_h = 1.3 \times 10^{-7}$ as given above. This gives $y \approx 1.9 \times 10^{-4}$. Then pH ≈ 10.3. Right away we see that the hydrogen ion concentration is small. The remaining concentration of HPO_4^{2-} is, to 2 SF, 0.068 M. We can check the other equations. It is immediately obvious that $[H^+] << [OH^-]$, and $[H^+]$ can be neglected. The same is true of the hydrolysis of $H_2PO_4^-$, and (slightly closer call) the further ionization of HPO_4^{2-}. We could calculate all these from the equations given above plus K_{a3}, with hydrolysis of $H_2PO_4^-$, completely negligible at this pH. We leave completion of the calculation as an exercise. We had 10.00 g Na_2HPO_4 to begin with; does this entitle us to 4 SF? What other quantities might be limiting — as an exercise, try the problem with initial concentration of $Na_2HPO_4 = 0.21$ M, triple what we just had. Does this change any of the approximations we made?

Summary: There are a number of equilibria that are specific to acids and bases, including K_w, as well as hydrolysis, buffer solutions, and acid-base titrations (there are other titrations, but the acid-base titrations are a particular class). Some of these equilibria are quite complex, and can only be solved by observing which quantities can be neglected.

Problems:

1) A solution of HCl is 0.23 M. What is the concentration of OH^-?

> *Estimate*: HCl is a strong acid, so it ionizes completely. Since the product $[H^+][OH^-] = 10^{-14}$, and in this case $1 > [H^+] > 0.1$ M, therefore $10^{-14} < [OH^+] < 10^{-13}$ M.
> *Answer*: $[OH^-] = 1.00 \times 10^{-14}/[0.23] = 4.3 \times 10^{-14}$.

2) Find the pH of a solution of 0.0163 M HCl.

> *Estimate*: This means that $[H^+] > 0.01$ M; $-\log_{10}0.01 = 2$, so the pH < 2
>
> *Answer*: pH $= -\log_{10}[0.0163] = 1.79$

3) Consider the following buffer: Mix 100.0 mL of 0.1000 M HAc, a weak acid with a K_a of 1.75×10^{-5}, with 100.0 mL of 0.0750 M of NaOH, a strong base. Find the pH.

> *Estimate*: Not all the HAc is neutralized: There are 0.1 L of 0.1 M HAc, so 0.0100 moles total H^+ available, even if not all is ionized. There are similarly 0.0075 moles of OH^-, which essentially all reacts with HAc. The ratio of unreacted HAc to Ac^- is 1:3.
>
> $$K_a = [H^+][Ac^-]/[HAc] = [H^+] \times 3 = 1.75 \times 10^{-5}$$
>
> So pH is more than 5. Here we need the extra SF in amount of HCl, because we are subtracting a number that is almost equal to the number it is being subtracted from.
>
> *Answer*: The estimate is almost the answer: $[H^+] = (1.75/3) \times 10^{-5}$, so pH $= 5.23$

COMMENT: As soon as we see that the amount of Ac^- is roughly the same (a factor of 3 is close enough) we should see that the final answer must be somewhere around the pK of HAc, in this case around 5. The estimate above already finds the actual ratio, so it is most of the way to the actual solution.

4) Suppose we have one liter of 0.10 M formic acid, $K_a = 1.8 \times 10^{-4}$, We want to add enough HCl from the vapor phase to lower the pH of the solution by 1 unit. Assume the change in volume of solution is negligible (a not completely unreasonable approximation). Find the volume of HCl vapor that must be added at STP to do this.

> *Estimate*: To lower pH by 1 unit, the amount of $[H^+]$ must be increased by a factor of 10. The acid pK is of the order of 10^{-4}, so the pH is between 2 and

3 (remember the solution is only 0.1 M, so we have less than a mole of formic acid). We would have to add enough HCl to get the pH around 1, so for 1 L, we need something a little less than 0.1 mole, or, at STP, a little less than 2 L of $HCl_{(v)}$.

Answer: We have to first find the starting pH, then the amount of acid we need to have the pH one unit below the starting value. The acid constant gives

$$K_a = [H^+][CHO_2^-]/([HCO_2H]_o - y)$$

$[HCO_2H]_0$ is the initial concentration of $[HCO_2H]$. And let $y = [H^+] = [CHO_2^-]$, so $0.1\ K_a = y^2$ to a first approximation

$$[H^+] = 4.2 \times 10^{-3} = 0.0042\ M, \text{ so we need}$$
$$0.0042 \text{ moles in 1 L}$$

The first approximation assumes $y \ll [HCO_2H]_o$ or $.0042 \ll 0.1$. This may not be an adequate approximation to enough SF, although it won't be off by terribly much. We could have guessed this from the ratio of initial concentration to pK, which is not larger by enough for the ionized fraction to be negligible compared to the initial concentration, something that we could have seen if we had just looked at the values we were given. Let's go back and solve the full quadratic equation, and we get $y = 0.00416\ M$. As the K_a value was only given to two SF, we could have lived with the initial approximation.

$$\text{So pH} = 2.38$$

We therefore need enough HCl to lower the pH to 1.38, which corresponds to $[H^+] = 0.041$, so in 1 L we need 0.041 moles, ten times the original concentration. If we start by assuming all the hydrogen ion comes from HCl, 0.041 moles, at 22.4 L molar volume = 0.92 L. However, not all the H^+ comes from the HCl. We should check the contribution of the $HCHO_2$ at the pH of the solution when the HCl has been added:

$$K_a = [H^+][CHO_2^-]/[HCO_2H]$$

$$1.8 \times 10^{-4} = 0.041[CHO_2^-]/[HCHO_2]$$

$$[CHO_2^-]/[HCO_2H] = 0.0044$$

This means that very little of the HCO_2H has ionized — it will contribute only about 1% of the total hydrogen ion. To two SF, there is no effect on the amount of HCl.

COMMENT: This is a problem that combines concepts from two chapters. The world is not built in separate chapters; while this is a somewhat artificial example, it is appropriate to be concerned with more than one set of ideas in a single question. We could also have guessed that the HCl would provide essentially all the $[H^+]$, as adding 10x the acid would suppress the ionization of the weak acid, which was already providing only 10% of the $[H^+]$.

5) Start with 100.0 mL of 1.00×10^{-3} M n-hexanoic acid (a carboxylic acid, also called caproic acid); n-hexanoic acid has $K_a = 1.8 \times 10^{-5}$ a) Find the initial pH. b) the solution is allowed to stand until half the water has evaporated; the vapor pressure of the acid is negligible — none of it evaporates. Find the final pH after evaporation.

Estimate: With a fairly dilute solution, where the K_a is much smaller than the concentration, the degree of dissociation is very small. Here the K_a is about two orders of magnitude smaller than the concentration, so we can guess that maybe 10% will dissociate. With the concentration doubling after the evaporation of half the water, we should expect a lower percentage of dissociation, so with a higher concentration of acid, there will be a slightly lower pH. If 10^{-4} moles of $[H^+]$ is present in 0.1 L, we would have a pH somewhat greater than 3. The more concentrated solution would have a pH about 0.1 lower, and with less dissociation.

Answer: $K_a = [H^+][A^-]/[HA] = [H^+][A^-]/([HA^\circ] - [H^+])$; usually we ignore the $[H^+]$ in the denominator on the grounds that $[HA^\circ[>>[H^+]$; in the preceding problem this would have been an adequate approximation. However, here degree of dissociation is not so small, so we should keep the $[H^+]$ in the denominator, leaving a full quadratic equation. Solving, we get

$$[H^+] = 1.35 \times 10^{-4} \text{ M, pH} = 3.87$$

This is about 1/8 of the starting concentration, so neglecting $[H^+]$ would not have been justified. If we had done this, the pH would be different by about 0.05 units. With twice the concentration (same amount of acid, half as much water), $[H^+] = 1.8 \times 10^{-5}$ M, pH $= 3.74$. The degree of dissociation is about 0.7 of what it was in the more dilute solution.

COMMENT: There is a fair amount of arithmetic here, but we can check that the answers are reasonable without doing the arithmetic that it took to get them. Obviously, there are instant limits: if all the acid dissociated, the $[H^+]$ would be 10^{-3} M. We can reasonably expect that the $[H^+]$ is somewhere near the square root of the initial concentration times K_a or, in this case, $[H^+] \approx 10^{-4}$. It makes sense to estimate where the answer should lie; if you see an answer that is an order of magnitude or more away from anything reasonable, most likely it is unreasonable. If you find the $[H^+] >$ the initial concentration of the acid, there is an obvious error. If you find the pH > 4, it is at least worth a very careful check; in fact, pH > 4 almost has to be wrong — try to figure out why — if it is >5 it is almost certainly wrong. This kind of consideration can come before the estimate, to make sure the estimate itself is not far off.

14

Chemical Kinetics

In determining what happens in the world, things can change because they are driven most strongly — as in having the maximum change in free energy, as we discussed in the chapter on thermodynamics — so a rolling stone, or a reaction, might find a minimum all the way at the bottom of the free energy hill. However, there may be a steeper path at a junction that takes the stone more quickly downhill, even if the end of this path is higher than the bottom of the hill. The analogy is to a reaction that winds up with a product that has a higher free energy than the most favored (lowest free energy) product, but is formed more quickly. If the reactants are used up in the faster reaction, that is the product that will be formed.

In a chemical reaction, there is a free energy path involved, from the reactants to the products. There is a maximum free energy between the reactant and product that forms a barrier. In a chemical reaction, usually a bond must be broken; this costs energy, and this energy must be added to allow reaction; then a new bond may form, and the free energy drops. Overall if there is a net decrease in free energy, the products are favored, but on the way the barrier, with the increased free energy, must be overcome. A broken bond, or partially broken bond, may not be the only reason that extra free energy is required, but it is perhaps the most common reason. In any case, the barrier must be overcome for the reaction to occur. We are not surprised that the rate at which the barrier is overcome is proportional to $\exp(-\Delta G^{\ddagger}/RT)$, where $\Delta G^{\ddagger}$ is the *free energy of activation*, and RT, as usual, is the thermal energy. The ratio of the size of the barrier to thermal energy determines the rate; a high barrier means a slow reaction. As the temperature increases, the rate rises in accord with the exponential. Define rate as the number of moles formed, or else reactant moles used up, *per unit time*. Let A be the proportionality constant, where we write rate as

$$\text{Rate} = A \exp(-\Delta G^{\ddagger}/RT)$$

If the rate is in moles per second, these must be the units of A as well. The exponent is dimensionless.

If the log rate is plotted against $1/T$, one gets a straight line, with slope $-\Delta G^{\ddagger}/R$. The constant A is sometimes thought of as an attempt rate to cross the barrier, and is generally of the order of a molecular vibration, perhaps 10^{12} s^{-1}. If you consider a low frequency vibration, which is essentially classical as it is low enough frequency that quantum effects are lost, $A \approx k_B T/h \approx 6 \times 10^{12}$ s^{-1}. If the activation energy is 25 kJ and the temperature rises from 300 K to 310 K, the rate increases by the ratio $\exp(-25000/R \times 310)/\exp(-25000/R \times 300) = 1.39$, (the activation energy is written here in Joules, not kilojoules, to match R, which is 8.31 J K^{-1} $mole^{-1}$) so the rate is almost 40% faster for a 10° rise in temperature. The extent of the speedup with temperature depends on activation energy. Larger activation energy means a greater temperature effect. In this example, the activation energy is not too large, 10 $k_B T$, so if $A = 10^{12}$ s^{-1}, then the rate is very fast, of the order of 10^8 s^{-1}.

There is in general a relation between the $\Delta G^{\ddagger}$ and the ΔG° of the overall reaction that appears in K_{eq}; Consider the reverse reaction. If the $\Delta G^{\ddagger}$ for the forward reaction is labelled $\Delta G_f^{\ddagger}$ then the reverse reaction (the reaction in which the previous products are the reactants and the previous reactants are the products) has its $\Delta G^{\ddagger}$, say $\Delta G_r^{\ddagger}$. Then for the overall reaction

$$\Delta G^{\circ} = \Delta G_f^{\ddagger} - \Delta G_r^{\ddagger}$$

The difference between the reactants and products is the same, whichever is reactant, or whichever is product — only the sign is reversed. One way you are going uphill for the overall reaction, the other way downhill.

Catalysts: If the reaction is too slow to be practical, it may be possible to speed it by adding a substance that offers a different reaction path; if this has a smaller barrier, the reaction, in both directions, will be faster. The free energy minima on both sides of the hill don't change. Such a substance is called a *catalyst*, and we will discuss these further along. For now, see Fig. 14-1, which shows a schematic diagram of a reaction along a "reaction path"

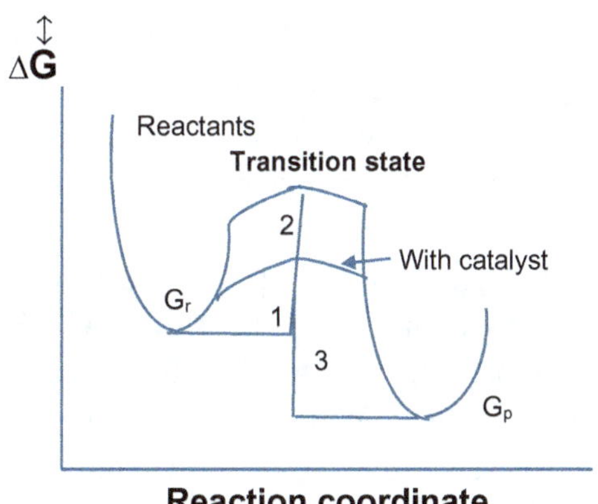

Fig. 14-1. Free energy diagram for a reaction, uncatalyzed and catalyzed. The overall free energy for the reaction is the difference between the minima of reactants and products, $\Delta G(\text{reaction}) = G_p - G_r$ and is in this case <0, so products are favored. The activation energies are as follows (not labeled as activation energies in the diagram, as that is too crowded.) Uncatalyzed: forward activation energy = 1 + 2, reverse activation energy = 3 + 2. The reaction energy is the difference between forward and reverse activation energies, 1 − 3. Catalyzed: forward activation energy = 1 reverse activation energy equals 3. The catalyst does nothing to the overall reaction energy, or to the equilibrium constant, but because energy 1 << energy 1 + 2, it speeds the reaction greatly. The increase in rate with the catalyst is A exp $(G^{\ddagger}(2)/RT)$; the energy required to get over the barrier is smaller by $G^{\ddagger}(2)$. The catalyst is not changed in the reaction.

(different for every reaction), and also shows how the transition from reactants to products could follow a different path with a smaller barrier, in the presence of a catalyst; the catalyst is not changed in the reaction. It is not a reactant.

What is true for the free energy is also true for the other thermodynamic functions of state, including in particular $\Delta H°$ and $\Delta S°$. If any of these is positive going forward, it will be negative going backward. Recall exothermic, endothermic reactions from the thermodynamics chapter — if forward is exothermic, reverse is endothermic.

Many reactions are dependent on breaking a chemical bond, at least partially, before a new bond can form, so it is not unusual to have activation energies in the general neighborhood of 100 to 200 kJ, which is an appreciable fraction of a bond energy. If an activation energy is about 100 kJ, increasing the temperature from 300 to 310 K will come close to tripling the rate. From 600 to 610 K the effect is smaller, more like increasing the rate around 1.5 to 2 times. You may wish to plug some numbers into a calculator to get a feel for the extent to which rates increase with temperature; the calculation is the same as the one we did with 300 to 310 K with 25 kJ activation energy. If the activation energy is larger, the temperature effect will be larger in any given temperature range, and the absolute value of the rate constant correspondingly smaller. In the example above, in which the activation energy was only 25 kJ, a 10 K increase in temperature produced only a 40% increase in rate near 300 K. This will give you some sense of what range of answers to expect for many of the kinetic problems.

There may be competing reactions, and the one with the lower $\Delta G^{\ddagger}$ will be faster, so the products of that reaction will dominate, even if the $\Delta G°$ for another reaction starting with the same reactants is more negative at equilibrium, with different reaction products. If there is a competing reaction that, uncatalyzed, is 10 times faster, but with the catalyst the previously slower reaction is speeded by a factor of 3000, the previously slower reaction is now 300 times faster. The products in this case are those of the catalyzed reaction instead of the reaction that would have predominated without the catalyst. Also, the reaction completes in 1/300 the time that the other reaction would have required. Rates for reactions can vary from fractions of a nanosecond to times long compared to the age of the universe — naturally we will not observe the latter.

The absolute value of the rate is also determined by the concentrations of the reactants — if the reaction depends on two molecules meeting, this will happen more frequently if there are more molecules.

If the reaction gives off heat (is exothermic), the temperature may rise, and the reaction would then speed up accordingly. In this case, the mechanism is different, and we won't discuss it further here, except to say that a barrier must be overcome.

In biological reactions, the catalysts are enzymes, which are proteins. A mutated enzyme that cannot speed the right reaction can be fatal, and often is. Protein structures can undergo hinge motions, so a protein that can open like a clamshell can trap two reactants, close until the product forms, then open, releasing the product; this can be repeated many times per second, and offers an example of an alternate means of speeding a reaction without actually changing the barrier. However, changing the barrier is most common. The clamshell conformational change can speed the reaction as well.

In biology, catalysis is crucial. Failure to produce a needed compound, or to clear a waste product, can be lethal. There are a number of hereditary diseases in which a mutation disables an enzyme. Biological reaction networks are complex, and disruption of a complicated set of reactions because of the failure of a single step can be extremely serious, which is why such diseases are often lethal.

Reaction order: Reactions typically go through *intermediates*, products that form transiently and then are converted into the final product in another step, or steps. Sometimes such mechanisms are complex, and we cannot go through them in an introductory course. However, we can describe the reactions in terms of the **dependence of rate on concentration of reactants**. Some reactions are proportional to the first power of one reactant concentration; such reactions are called first order. Some reactions are proportional to the product of the concentrations of two reactants, or to the square of the concentration of one reactant. These are called second order. You can even have independence of concentration (zero order) or dependence on the third power of concentrations, naturally called third order. Because of intermediates, in which one step makes a product that is the reactant in a second step, with the steps being of different order, or otherwise affected by relative rates, you might also find the rate to be a complicated expression, not of any simple order. However, first and second order are common. If we are discussing a single reaction step and find experimentally that it is proportional to the product of the concentration of some reactant A with the concentration of another reactant B, we would very possibly guess that the reaction step depends on the collision of an A with a B — but this is not certain — see the paragraph on molecularity below. Studying the concentration dependence of a reaction, or, if a step can be isolated, the concentration dependence of a step, is one way to work out a mechanism for that step. Working out the consequences of a mechanism mathematically is a great deal easier if you have some elementary calculus, so we omit it here. However, we can remark, for those who have begun calculus, that we defined rate as rate of change of a concentration; for rates, it is easiest to work with derivatives, and to turn these into concentrations as a function of time, and to then integrate the differentials that you get. The units of the proportionality constant to the concentrations are important. We can define a *rate constant*, such that

Rate = k_R [A][B] (second order). Or **rate = k_R[A]** (first order)

Or, possibly other forms — k_R is the proportionality constant in the rate expression, that is, the rate constant, and the rate expression need not be first or second order.

The rate must be of the form concentration/time, so with the molarity of the reactants appearing to the zeroth, first, second, or third power, the rate constant k_R must have the units of $[R]^{-(n-1)}\,s^{-1}$, where [R] is the concentration of reactant R in M L^{-1} and n is reaction order, so that we always have M $L^{-1}\,s^{-1}$ for the units of the rate at which a reactant is used up, or a product appears, in the equation:

$$\text{rate} = -k_R\,[R]^n$$

If you know a little calculus, you can write rate as d[R]/dt, where [R] is the concentration of a reactant in M L^{-1}. If it is a first order reaction (n = 1) you can integrate to get the exponential form of the rate law, and then taking the log, you have

$$\log([R]/[R_o]) = -k_r t,$$

where $[R_o]$ is the initial (t = 0) concentration of R. If you don't know calculus at all, just accept this result. By plotting $\log([R]/[R_o])$ vs t, you get a straight line. You can also integrate the second order (n = 2) case; a plot of 1/[R] vs t gives a straight line, if we assume that it is second order because of proportionality to $[R]^2$.

$$1/[R] - 1[R_o] = k_r t$$

If you have a first or second order reaction, one of the two gives a straight line, from which you can tell whether you have a first or second order reaction; there is no guarantee that a reaction is simply first or second order, so you may not get a straight line either way; there may be an intermediate that completely changes the apparent order of the reaction, possibly to something that looks like a fractional power. The apparent order may even change as the concentrations change during the course of the reaction. Details have to wait for a more advanced course, not because of conceptual difficulties, but because the reaction mechanism can produce complications (but see the next paragraph). If you do get a straight line showing either first or second order kinetics, the appropriate slope gives the rate constant. It is similarly possible to find a plot giving a straight line that fits a zero or third order reaction, but these are relatively rare. The relation of the disappearance of reactant to the rate of production of the product depends on the stoichiometry of the reaction.

If the reaction has multiple steps, the overall rate at which reagent is used, or product appears, depends primarily on the slow step. If you are driving down a highway in which you pass four landmarks, and the first, second, and fourth take 5 minutes each, but step three has a traffic jam that takes 3 hours, your trip time will approximate the 3 hours, the slow step.

Molecularity: Because there can be multiple steps, and the order may be determined by a slow step, the order may give us the number of molecules that must come together, or prior to, in a single slow step; this is called the *molecularity* of the reaction. Each step may be different in this respect, and the overall order that is measured may depend on the slow step, or on a fast step that occurs prior to the slow step.

Fast and Slow Steps: the fact that some reactions proceed in more than one step is responsible for the fact that reactions do not need to be of integral order. Reaction rates for different steps are different, and the slow step takes up most of the time, so the overall rate is the rate of the slow step. If it is preceded by a fast step, and this fast step is reversible, it is safe to take the fast step reactants and products to be at equilibrium before the concentration of the product of this step is significantly depleted by the ongoing slow step. A fast step after a slow step is rarely important, as the product of the slow step is immediately (on the time scale of the slow step, thus of the entire reaction) converted to the final product.

Summary: The rate of reaction, that is, the amount by which concentration of a reactant or product changes per unit time, is very informative as to the mechanism by which the reaction occurs. There is a barrier to the transition from reactant to product, which can be expressed in terms of thermodynamic quantities; at the usual conditions of constant temperature and pressure this is the free energy term $\Delta G^{\ddagger}$, called the activation energy. This barrier can be lowered by a catalyst, which alters the mechanism, most often (not always) by providing a reaction path with a much lower barrier, so that it proceeds much faster. Raising the temperature also speeds a reaction, as the critical quantity is the ratio of the free energy of the barrier to the thermal energy. The concentration dependence of the rate can be informative as to the mechanism of the reaction. In biology reactions are connected in a complex web. The loss of an enzyme can lead to the effective loss of a reaction, as it slows so much that its products are unavailable for further steps; this can be lethal.

Problems:

1)   The reaction $O_3 + O \rightarrow 2\,O_2$ is first order in both O_3 (ozone) and O (atomic oxygen) (therefore second order overall). At 25°C the rate constant is 7.8×10^5 L mol^{-1} s^{-1}. When $[O_3] = 5.0 \times 10^{-3}$ M, the rate is measured to be 9.75×10^{-5} M L^{-1} s^{-1}. What is the concentration of atomic oxygen, O?

   Estimate: This is simple enough that all we can do is write down the equation and round off: rate $= k_r\,[O][O_3]$

(both are given as first order), so $[O] = \text{rate}/k_r[O_3]$; rounding to one SF, $[O] = 10^{-4}/(10^6 \times 10^{-2})$ or about 10^{-8}. *Answer*: plug in the actual values, to get $[O] = 2.5 \times 10^{-8}$. The rate constant limits the answer to 2 SF.

COMMENT: Make sure you get a plausible order of magnitude. Here the estimate only helps to make sure the answer is not wildly off. The concentration of a very unstable species like atomic oxygen can't be high. If you accidently wrote the rate as 9.75×10^5, which would make the answer ten orders of magnitude larger, it would be ridiculous on its face. In this case, your chemical intuition is all you have to go on to make sure you didn't do something silly — it is easy to punch in the wrong button on a calculator but it takes some sense of what is reasonable to reject the unreasonable. Another comment: Suppose $[O_3]$ were given as of order 10^3 instead of 10^{-3}. Would this be reasonable? There is a *critical point* above which there is no longer a distinction between liquid and vapor phase. For O_3 this temperature and pressure are $-12.1°C$ and 54.6 atm pressure. At 1000 atm and $25°C$ what would the O_3 be like?

2) Part 1: A third order reaction has rate $= k_e[A]^2[B]$. Under which of the following conditions does the greatest **decrease** in rate occur? How large is the decrease?

    a) [A] is halved
    b) [A] is doubled
    c) [B] is halved.
    d) [B] is doubled
    e) [A] and [B] are both halved
    f) [A] and [B] are both doubled

    Estimate and answer: Obviously b), d). and f) go in the wrong direction, and the more either [A] or [B] decrease, the more the rate decreases, so e) is the answer without much thought, and the rate decreases by 2^3 or 8.

Part 2: There actually are a few third order reactions, including the gas phase reaction

$$2NO_{(g)} + O_{2(g)} \rightarrow 2NO_{2(g)}$$

Here NO plays the role of A in Part 1, and O_2 the role of B. We could try various combinations of concentrations to see which gives the fastest rate, which would be the inverse of Part 1. However, instead, find the $[NO_2]$ at t = 300 s, if the initial concentration of $[O_2]$ is 0.04 M (about that in the gas phase at atmospheric pressure and 25°C, if it is pure O_2, not air); given, at t = 100 s $[NO] = 1.0 \times 10^{-4}$ M L^{-1} and at 200 s $[NO] = 0.25 \times 10^{-4}$ M L^{-1}.

Estimate: First of all, let us write down the rate expression:

$$\text{Rate} = \mathbf{k}_r[NO]^2[O_2]$$

We can do this because we are told that it is a third order reaction, and with 2 NO reacting with 1 O_2, this is what makes sense. To check this, note that [NO] decreases 4-fold when the time doubles, which helps confirm the dependence on $[NO]^2$. The rate at which NO_2 forms is the same as the rate at which NO disappears, as required by the stoichiometry of the overall reaction.

We are asked to find the [NO] at 300 s. $[O_2]$ is nearly constant, as $[NO] \ll [O_2]$. We might reasonably expect something of the order of 10^{-4} M L^{-1}, since this is about the order of magnitude of the change in the reactant. Second, instead of treating this as a third order problem, treat it as second order in [NO]. That is, the reaction might as well be second order in NO, while oxygen is treated as a constant.

Then $1/[NO]_0 - 1/[NO] = -\mathbf{k}_r t$

A plot of $1/[NO]$ against time gives a straight line, with slope $\mathbf{k}_r$, which in this case includes the essentially constant concentration of oxygen.

Answer: Use the values of [NO] at 100 s and at 200 s to get the $\mathbf{k}_r$, Then plug in 300 s.

$$1/1.0 \times 10^{-4} - 1/0.25 \times 10^{-4} = -\mathbf{k}_r t$$

Since the value at $t = 100$ s is taken as the initial value here, we have the time from 100 s to 200 s, making the reaction time $t = 100$ s, so $\mathbf{k}_r[\mathbf{O_2}] = 3 \times 10^2$ L mol^{-1} s^{-1}. Then at 300 s,

$$1/10^{-4} - 1/[NO]_{300} = -3 \times 10^2 \times 200.$$ This [NO] is at $t = 300$ s, or 200 s after the given initial value

$$\text{So } [NO]_{300} = 0.14 \times 10^{-4} \text{ mol L}^{-1}$$

COMMENT: We can extract the concentration of O_2, giving $\mathbf{k}_r = 7.5 \times 10^3$ L^{2}mol^{-2}s^{-1}

3) For the same reaction as in Problem 2, we are given the following data, this time with $[O_2]$ not in excess, and at a much lower temperature, so the reaction is slower:

Table 14-1.   [NO], $[O_2]$ concentrations in [**M**]

[NO]	[O₂]	rate(NO₂ production), 10^{-12} M s^{-1}
0.001	0.001	15
0.001	0.002	30
0.001	0.003	45
0.002	0.003	180
0.003	0.003	405

Part 1: find the rate of appearance of $[NO_2]$ when $[NO]$ = $[O_2]$ = 0.01 M

Part 2: find the rate constant k_r

Estimate: First of all we can see that the rate is slow, given in units of 10^{-12} M s^{-1}, so we expect k_r to be appreciably less than one. We can look at the Table, and see immediately that if $[NO]$ is constant, the rate increases proportionally to the concentration of oxygen — that is, it is first order in $[O_2]$. From the last three lines, with $[O_2]$ constant, it is obviously second order in $[NO]$. Overall, therefore this is a third order reaction; we don't need problem 2 to see this. We are asked to find the rate of production of $[NO_2]$ at a concentration that is out of the range of the Table of data, but we are therefore implicitly to assume that the mechanism, or at least the order, and rate constant, remains the same. If we take the second line of the Table, in which both reactant concentrations divide evenly into 0.01 **M**, $[NO]$ = 0.1 times the given concentration, $[O_2]$ = 0.2 times the given concentration, and the former has to be squared as it is second order with respect to $[NO]$, so altogether the result should be 500 times the rate of production of $[NO_2]$ in that line, so 15000×10^{-12} M s^{-1}, and this is not only the estimate but the answer. For k_r, we observe that with concentrations of 10^{-3} and a third order overall reaction, we get around 10×10^{-12} as a rate, so we expect something of the order of 0.01.

Answer: The estimate already gave us the answer to Part 1). For Part 2. instead of looking at the order of magnitude of the rate, we can look at the actual value, and find k_r = 0.015. This can be checked by using other lines in the Table. For example, use line 4.

$$Rate = 180 \times 10^{-12} = 0.015 \times 0.002^2 \times 0.003?$$

And this checks.

COMMENT: This is not an air pollution problem. The concentration of oxygen is far too low for breathable air, and the concentrations of nitrogen oxides too high to be compatible with breathable air.

4) Nitrogen oxides are not the only serious air pollutants. The reaction

$$C_2H_4 + O_3 \rightarrow 2\,CH_2O + \tfrac{1}{2}\,O_2$$

C_2H_4 is ethylene, CH_2O is formaldehyde. Air is considered severely polluted with $[O_3] > 5 \times 10^{-8}$ M and $[CH_2O] > 1 \times 10^{-8}$ M. We have the following data (Table 14-2), with **initial concentrations of ozone and acetylene given in units of 10^{-8} molar:**

A. Find the order with respect to each reactant, and find the rate constant

B. At the concentrations in the second line, how long until CH_2O reaches its danger level, if it starts from zero?

   Estimate: A) For O_3, combining lines 1 and 3, we see that tripling the concentration leads to triple the rate of formation of CH_2O, with constant C_2H_4, so the reaction is first order in O_3. Combining lines 1 and 2 shows a four-fold increase in rate with both reactants doubled. One factor of 2 we now know comes from O_3, so doubling C_2H_4 doubles the rate, and the reaction is also first order in C_2H_4. The reactant concentrations are of order 10^{-7} (10×10^{-8}) and 10^{-8}, respectively, and the rate of order 10^{-12}, so k_r must be of order 10^3.

Table 14-2.

$[O_3]$	$[C_2H_4]$	rate $[CH_2O]$ 10^{-12} Mole L^{-1} s^{-1}
5.0	1.0	1.0
10.0	2.0	4.0
15.0	1.0	3.0

B) The rate is of order $10^3 \times 10^{-15}$, and the danger level 10^{-8}, so if we guess 10^4 s we should not be too far off.

Answer: A) The estimate is the answer for the first part. To find the actual rate constant, try line 1 of the Table:

$$\text{rate} = 1.0 \times 10^{-12} = k_r \times 5.0 \times 10^{-8} \times 1.0 \times 10^{-8}, \text{ so}$$

$$k_r = 1.0 \times 10^{-12}/5.0 \times 10^{-16} = 2.0 \times 10^3$$

B) Here, we have the reactant concentrations constant, so the rate is constant at 1.0×10^{-12} [M] s^{-1}, and it will take just 1.0×10^4 s to reach 10^{-8} [M]. As long as the reactants are not used up, a reasonable assumption in air pollution, the reaction rate would be constant.

COMMENT: Obviously this data has been rounded off. However, the data, and the danger, are real enough. Air pollution varies from place to place even over a fairly local area, so even one SF is an average at best.

5) Esterification: Acetic acid reacts with ethanol to produce an ester, ethyl acetate. The reaction is reversible, and reaches equilibrium (see Chapter 10, problem 4).

$$CH_3CH_2OH + CH_3COOH \leftrightarrow CH_3COOC_2H_5 + H_2O$$

K_{eq} at 25°C is 2.92. The forward rate constant at 25°C is 2.38×10^{-4} L $mol^{-1}s^{-1}$. What is the rate constant for the reverse reaction (the hydrolysis of the ester)?

Estimate: There is equilibrium, so the equilibrium constant is the ratio of the rates of the forward and reverse reactions. Forward is of the order of 10^{-4}, with equilibrium constant >1 but <10, so reverse is slower than forward. The order of magnitude for the reverse reaction must be around 10^{-5}.

Answer: $K_{eq} = k_r(\text{forward})/k_r(\text{reverse}) = 2.92$; then $k_r(\text{reverse}) = k_r(\text{forward})/K_{eq} \approx 2.38 \times 10^{-4}/ 2.92$. To get

3 SF, just plug in the actual values: $k_r = 8.15 \times 10^{-5}$ L mol^{-1}s^{-1}.

6) For the reaction $A_2 + B_2 \rightarrow$ products, the measured rate is

$$\text{Rate} = k_r\, [A_2][B_2]^{1/2}$$

The following mechanism is proposed to give this dependence on the reactant concentrations:

(1) $B_2 \rightarrow 2B$ fast, reversible
(2) $A_2 + B \rightarrow$ product   slow

Would this mechanism account for the observed rate law?

Estimate: Here there are no numbers — this problem is included to illustrate fast and slow reactions and the rate law that results from a simple multistep (here, two step) reaction.

Answer: We can assume that the first step reaches equilibrium before the slow step affects concentrations.

$$K_{eq}\,(1) = [B]^2/[B_2]$$

So $[B] = (K_{eq}\,(1) \times [B_2])^{1/2}$
Then, for the slow step,

$$\text{Rate} = [A_2][B] = [A_2]\,(K_{eq}(1) \times [B_2])^{1/2}$$

This does give the same form as is found experimentally, so it is a plausible mechanism, and deserves further investigation.

COMMENT: The fact that it gives the same answer as experiment keeps it as a possibility, even a strong possibility, and a justification for further investigation. For example, this suggests that the monomer, B, should be present in substantial quantity. Does this show up in spectra of the reaction mixture? The light absorption can in principle give the concentration of B as a function of time; this could be used to confirm, or rule out, this proposed mechanism. If the light absorption turns out to be an unsuitable analytical method (there are many ways this can be true), other analytical methods might be found.

7)   The activation energy for the decomposition of acetaldehyde is 188 kJ mol^{-1}. The rate is 2.45×10^{-8} L mol^{-1} s^{-1} at 500 K.

a)   At what temperature will the rate constant double?
b)   What will the rate constant be at 600 K?

Estimate: This is a fairly large activation energy, so looking for doubling in about 10 K in the 500 K range is probably going to be roughly correct; make an initial guess of 510. The rate at 600 K may be fairly large, relative to the rate at 500 K as the 188 kJ activation energy is large. This depends on our having a feel for these numbers, rather than doing any calculations.

Answer: a) Find the temperature at which the rate is 4.90×10^{-8} L mol^{-1} s^{-1}

We have not found the pre-exponential term. Either we set up the ratio of rates to cancel it, or solve for it at 500 K and as in the other method, assume the pre-exponential term to be constant, which is the same assumption we make in cancelling it. Set up the ratio:

Rate(T)/Rate(500) = 2 = exp(−188000/8.31*T)/ exp(−188000/8.31*500)

With the exponentials, it is better to take the natural logs

ln (2) + (−188000/8.31*500) = −188000/8.31*T

$$T = 507.8 \text{ K}$$

This is not so much a 4 SF answer, as a 2 SF answer, since the difference from the starting 500 K is given to two SF; the 500 is given as exact.

c)   Here we know the temperature, and need to find the rate constant:

$$\text{Rate (600)/Rate(500)} = \exp(-188000/8.31*600)/\exp(-188000/8.31*500)$$

$$\ln(\text{Rate(600)}) = \ln(\text{rate(500)}) -188000/8.31*600 -(-188000/8.31*500)$$

$$\ln(\text{Rate(600)/Rate(500)}) = 7.5411$$

$$\text{Rate(600)} = 4.79 \times 10^{-5}$$

The ratio of the rates at 600 and 500 is 1880 (3 SF)

COMMENT: The large activation energy is responsible for the fact that the doubling temperature was only about 8 K, and this is consistent with the large ratio in the rates of the 500 to 600 K reactions. There are more than 12 doublings in that temperature interval. $2^{12} = 4096$; however, as we metioned in the text, the higher temperature shows a smaller temperature effect. That is, as the temperature gets higher, the doubling range gets larger.

15

Electrochemistry

A great deal of chemistry happens at electrodes, in which oxidation occurs at one electrode, and reduction at the other. Batteries are electrochemical cells that are significantly out of equilibrium, and the energy they give off as they approach equilibrium is used to produce an electric current in the external circuit. When the battery reaches equilibrium, that is, its minimum free energy, it is dead. Many batteries can be recharged by putting in energy in the form of an electric current, reversing the reaction so that they store the energy, with the reaction products converted back to the reactants. The energy that has been put in is stored as the higher free energy of the reactants. When they are again connected to an electrical circuit, the forward reaction starts; that allows the energy stored in the out-of-equilibrium form to be used as an electrical current that can do work.

Units: The fundamental S.I. electrical unit is the *Ampere* (A); define Coulomb by making 1 A a current of one Coulomb (C) per second (Coulomb is therefore 1 ampere-second). One electrical charge on an electron or proton is 1.602×10^{-19}C; thus, one mole of charge is $6.0223 \times 10^{23} \times 1.6022 \times 10^{-19}$ C $= 96,531$ C $= 1$ *Faraday* of charge, which is written $\mathcal{F}$. The other major quantity we need is energy, and 1 C $\times$ 1 Volt (V) $= 1$ J; Joules and Coulombs having been defined in terms of fundamental units, this relation defines Volt. The unit of electrical potential is the Volt. A voltage difference is the driving force for electrical current. Dropping 1 charge through 1 V gives 1 eV, so 1 eV $= 1.602 \times 10^{-19}$J, a relation we already met in Chapter 7. Dropping 1 $\mathcal{F}$ of charge through 1 V = 96.531 kJ. Therefore, if we know how much charge is stored in a battery, at what voltage, we know how much energy we have. It is useful to keep these numbers in mind (at least rounded off) as we discuss electrochemical cells, and their uses. Charges $\times$ voltage = energy, and this is also the contribution of the electrical energy to the free energy.

Review Before we begin considering chemical reactions at electrodes, we should consider the magnitude of the electrical current. To repeat the key definition, *Amperes* is one of the seven fundamental units of the SI system, and with it we define Coulombs, the unit of charge, from the relation (Q is charge, measured in Coulombs); repeating the definition of Coulombs

$$Q = A * s \text{ (ampere seconds), with one Coulomb (C) equal to 1 A*s}$$

In chemical lab experiments, we might consider current in mA (milliamperes, or 0.001 A), a current of 5 mA for 10 s gives 0.05 C. Biological currents are often in the pA range (picoamp = 10^{-12} A). A 100 pA current for 10 ms (1 ms $=10^{-3}$ s) range gives 100 pA $\times 10^{-2}$ s $= 10^{-12}$ C, or roughly $10^{-12}/1.60 \times 10^{-19} \approx 6 \times 10^{6}$ charges; this is about the amount that might go through a biological membrane protein when it is in a conducting state that lasts for 10^{-2} s. Now that we have some sense of the magnitude of small currents (obviously if you want to run a large machine, you would be dealing with much larger currents, but in chemistry and biology we usually are not running anything very large on a macro machine scale). Current of course is a flow of charge, so it must be going from some place with a higher potential (voltage) to a place with a lower potential. So far we have said "higher" and "lower" without paying attention to sign; current is, by convention, positive in the direction of motion of positive charge, even if the actual charge carriers are electrons, as in a metal; current is positive in the opposite direction from the direction electrons move. There is always a driving force associated with any flow, and, if there is a flow, it is necessarily out of equilibrium; as we noted earlier, for batteries, as for any system, equilibrium is death — nothing changes, nothing moves. With this understood, we can proceed to consider the equilibria, as well as the out of equilibrium processes that define electrochemistry: the chemical reactions, and the potentials, that are involved in electrochemistry.

Reduction-Oxidation: Redox Reactions

A chemical reaction can have one, often two, occasionally three electrons exchanged, and for complex ions, in principle, at least, up to 7. Thus, per mole of reaction, $\Delta G = -z \, \mathcal{E}\mathcal{F}$, or we can write $\mathcal{E} = -\Delta G/z\mathcal{F}$, where $\mathcal{E}$ is the voltage, sometimes called the electromotive force; the bigger the voltage, the bigger the "force" driving the reaction forward (or backward, if the sign is reversed), and z is the number of electrons that are transferred in the reaction. If the reaction is to reach equilibrium, ΔG becomes $\Delta G°$, and $\mathcal{E}$ thus $\mathcal{E}°$.

There are two kinds of questions we can deal with when concerned with electrochemical cells: 1) how much current can we get, at what potential (measured in voltage) In other words, what is the energy we use/generate in this reaction. 2) Given an oxidation-reduction (**redox**) chemical reaction find the equilibrium concentrations, or else the voltage that can be generated. That is, a redox reaction is one in which electrons are exchanged so that the reaction involved leads to one reactant giving up one or more electrons, thus becoming more positively (or less negatively) charged, hence **oxidized**, while the other accepts electron(s). becoming **reduced** (less positive, or more negative).

In practical cases, we find that the specific nature of the electrodes matters, that equilibrium is rarely really achieved, and that in general one cannot simply write down a redox reaction, split it into two half-reactions (for half reaction definitions, see below), one that is an oxidation and one that is a reduction, and then expect that we can create a practical battery from this reaction. However, we must begin with the equilibrium reaction, which sets a kind of limit on what is possible. Practical matters that deal with the rate at which energy can be withdrawn from the battery, or the rate at which it can be charged, or the lifetime of the battery before some irreversible damage to the battery occurs, are left for engineering courses.

The generic form for the treatment of these questions can be summarized more or less (and without some significant details) as follows:

$$A + B \rightarrow A^{z+} + B^{z-}$$

where A has transferred z electrons to B, so that A is oxidized and B is reduced. This is written as though A and B start with no charge; however, they may have a charge to start, and the change in charge in reduction or oxidation is in that case from whatever the starting charge is. There can be complex reactions in which the nominal charge on a given atom changes by as much as 7 charges (formally, at least, according to our conventions for writing reactions — just how literally to take this is another matter) or there may be a chemical reaction in which the products are not simply the starting materials with a charge change but an entirely different molecule. Also, there may be water, and hydrogen (hydronium) and hydroxide ions, in the reaction. Cases like these are considered in the problems. However, simply to illustrate the underlying principle, we can omit all these possibilities.

The current is carried by both cations and anions. Cations have positive charge and move toward the cathode, which is negatively charged in the solution; anions, with negative charge, move toward the anode, which has positive charge. The quantity of charge, in Coulombs, tells how much reaction occurs. You can find out how many moles of reactant get reduced or oxidized by matching the amount of current to the number of moles that react.

For example, if you have 0.1 A for 2000 seconds, you have 200 C. It takes almost 100,000 C (96531 C) to make a mole, so 200 C makes almost 2 millimoles (.002 moles) reacting. Suppose you are reducing a metal, such as Cd^{2+} with 2 electrons. Then you have almost 0.001 moles of Cd^{2+} that is reduced. If you are asked to find the grams of Cd produced as metal, given the molecular mass of Cd (112.4) you would deposit $112.4 * 0.001 = 0.108$ g of Cd (you have 96531 C, not quite 100,000 — the exact answer depends on how many SF you attribute to the time and current).

Half reactions:

We need to introduce half reactions so that we can discuss the reactions clearly. In a half reaction we look at the oxidation and the reduction reactions separately. We write the reactions so that the electron(s) being transferred is (are) shown explicitly. To make an electrochemical cell, it is necessary to combine an oxidation and a reduction half reaction so that the electrons are not left unbalanced; there are no free electrons, and the half reactions are formal devices that allow us to do computations and solve problems. Let $\mathcal{E}^{\circ}_{1/2}$ be the reduction potential for a half reaction, which is written, by convention, as a reduction:

$$\text{Oxidation: A} \rightarrow \text{A}^{+} + \mathbf{e} \quad \text{voltage} = -\mathcal{E}^{\circ}_{1/2}(\text{A})$$

$$\text{Reduction: B} + \mathbf{e} \rightarrow \text{B}^{-} \quad \text{voltage} = \mathcal{E}^{\circ}_{1/2}(\text{B})$$

Because the convention is to write the half reactions as reductions, to get the value of $\mathcal{E}^{\circ}_{1/2}$ for the oxidation, subtract the $\mathcal{E}^{\circ}_{1/2}$ as you find it in a table of reduction potentials.

$$\text{Overall: A} + \text{B} \rightarrow \text{A}^{+} + \text{B}^{-} \quad \text{voltage} = \mathcal{E}^{\circ} = \mathcal{E}^{\circ}_{1/2}(\text{B}) - \mathcal{E}^{\circ}_{1/2}(\text{A})$$

In other words, we can combine an oxidation and a reduction half reaction to make the complete redox reaction, with the oxidation half reaction written with a minus sign. Here we use $\mathbf{e}$ to represent an electron (before we used q for an electronic charge, but here we mean an actual electron). One other note of caution: to combine the reactions, there must be no left over electrons, so if one half reaction transfers two electrons, and the other half reaction transfers only one, the latter must be multiplied by two so that the electrons balance. Electrons cannot be created or destroyed, so the number of moles must be matched accordingly. When you multiply the half reaction by two (or whatever), the voltage does not change — the voltage measures the "push" of the half reaction, in the sense of the bond strength, the energy of ionization, the hydration, and other contributions to the energy of the reactants and products, under ambient conditions, like temperature. As with other chemical reactions, there is an energy associated with the reaction, and if the number of moles reacting is multiplied by a factor, then the energy must be multiplied by the same factor. Energy is an extensive quantity, voltage is an intensive quantity (remember the definitions from the chapter on units).

The reaction equilibrium gives $\Delta \mathbf{G}^\circ$, but we have not accounted for what happens if we do not have standard conditions, and we have not accounted for the driving force from concentration differences. To find $\Delta \mathbf{G}$ we recall our discussion of equilibrium, where we saw that

$$\Delta \mathbf{G} = \Delta \mathbf{G}^\circ + RT \, ln \, K$$

so that we should expect $\Delta G = \Delta G^\circ + RT \, ln \, K$
$$= \Delta G^\circ + RT \, ln \, [Pr]/[Re]$$
where $[Pr]/[Re]$ is our usual ratio of products/reactants in concentration units. We can also go back and write this in terms of the potential by dividing by $z\mathcal{F}$:

$$\mathcal{E} = \mathcal{E}^\circ + RT/z\mathcal{F} \, ln \, [Pr]/[Re]$$

Having come this far, it is time to see what magnitudes are reasonable. To begin with, the range of reduction potentials is generally within a range of approximately 6 volts; the convention for defining this range is as follows: Since it is impossible to measure just one half reaction, it is necessary to choose one and make it the half reaction against which all others are measured. This half reaction, defined as exactly zero reduction potential, $\mathcal{E}^\circ_{1/2} = 0$, is, by convention,

$$2H_3O^+ + 2e \rightarrow 2H_2O + H_2;$$

This makes the range of reduction potentials approximately -3 to 3 V. Our 0 V half reaction is chosen to be roughly in the middle of the range of reduction potentials. Therefore, a very strong reducing agent, like $Li \rightarrow Li^+ + e$, might be around -3 V without thinking about it (it is actually -3.04 V; lithium really wants to be oxidized), and a strong oxidizing agent wants to be near $+3$ V (fluorine, which really wants to be reduced to F^-, has half-reaction potential 2.87 V).

If we are doing a calculation, and find an $\mathcal{E}^\circ_{1/2}$ outside the range -3 V to $+3$V, we can be reasonably sure we are wrong. There are more $\mathcal{E}^\circ$ values in the range -1.5 V to $+1.5$ V than outside, and if our chemical sense tells us that we have what should be a good reducing agent with $\mathcal{E}^\circ > 0$, or a good oxidizing agent with $\mathcal{E}^\circ < 0$, we should want to recheck if we get $\mathcal{E}^\circ_{1/2}$ in the wrong range. Overall, the equilibrium constant for a reaction will run in the direction that gives $\mathcal{E}^\circ > 0$, since $\Delta G^\circ = -z\,\mathcal{E}^\circ \mathcal{F}$ (note the minus sign) and the spontaneous direction of a reaction is that which has $\Delta G^\circ < 0$.

What about the effect of concentration? How much is $RT/z\mathcal{F}$? If $z = 1$, and $T = 298°C$ (standard temperature), then $RT/z\mathcal{F} = 25.6$ mV. If we switch from natural logs (ln) to $\log_{10}$ then we multiply by 2.303, which gives 59.0 mV for a factor of 10 in concentration ratio. In other words, if we have a large concentration ratio, we can expect to shift ΔG by not much more than 0.1 V, which amounts to $96{,}531 \times 0.1 \approx 10$ kJ. This is around $4\ k_B T$, and can shift a reaction ratio by around $\exp(4) \approx 50$, not negligible, even if not huge.

One thing we have not covered is the line notation for electrochemical cells. This is very useful as a notation, but adds little to physical understanding, and takes some time to learn, so we omit it here.

SUMMARY: Once we get used to the units, electrochemical reactions can be treated much like other reactions. However, there may be reactions at electrodes, thus introducing complications at surfaces, and inhomogeneity, not present in a chemical reaction that takes place in a single flask without phase boundaries, containing a reacting solution. Because the reactants and products of an electrochemical reaction have different oxidation states (more or less electrons) reactions can be written as the combination of two half reactions. The stoichiometric coefficients must make the number of electrons balance. There is a range of plausible voltages associated with the half reactions, and voltages must fall within the range which is possible. There is an energy associated with these reactions as with all others, and keeping in mind that voltage $\times$ charge is energy, we can do thermodynamics as usual with these reactions. The concentration dependence must be included in the calculation of ΔG.

Problems:

1) A current of 100 mA is passed through a solution of acid. For how long must the current run to produce one liter of H_2 at STP?

   Estimate: One liter at STP is <0.05 moles of H_2 (22.4 L > 20 L, so 1 L < 0.05 moles). However, the gas is H_2, so we need two electrons for each molecule. Therefore we need a little less than 0.1 mole of electrons. 100 mA $\approx 10^{-6}$ mole s^{-1}, so we need roughly 10^5 s, or a little more than a day.

   Answer: We need 1/22.4 moles of H_2 = 2/22.4 moles of electrons. Here they are being delivered at a rate of

0.1/96531 moles s^{-1}, so we need 2/22.4 moles/(0.1/96530) moles s^{-1} = 86200 s, or almost 24 hours.

2) In the reaction $Cd^{2+} + Fe_{(s)} \rightarrow Cd_{(s)} + Fe^{2+}$
find the ratio $[\,Cd^{2+}]/[Fe^{2+}]$ in the solution when equilibrium is reached at 298 K, so that there is no further reaction.

$$\text{Given: } Cd^{2+} + 2e \rightarrow Cd \quad \mathcal{E}°_{1/2} = -0.403 \text{ V}$$

$$Fe^{2+} + 2e \rightarrow Fe \quad \mathcal{E}°_{1/2} = -0.440 \text{ V}$$

Estimate: For equilibrium, we need $\Delta G = 0$. Here, the iron couple has the larger reduction potential, and if the ratio $[\,Cd^{2+}]/[Fe^{2+}]$ were equal to 1, the cell would produce a voltage of +0.037 V, with iron oxidized and cadmium reduced. We need to reduce this to zero. With a 2 e reaction, a ratio of 10 corresponds to about 30 mV. We therefore need $\log_{10}[\,Cd^{2+}]/[Fe^{2+}]$ to be around 1.2, so the ratio must be around 1/16. Without doing any calculations, we see that the direction of the reaction is forward as written — the iron is oxidized, and its potential must therefore be reversed, becoming positive. Therefore, there must be less Cd^{2+} and more Fe^{2+} to bring the voltage, thus the ΔG, to zero.

Answer: $\log_{10}[\,Cd^{2+}]/[Fe^{2+}] = -.037/.0295 = -1.254$.

$$[\,Cd^{2+}]/[Fe^{2+}] = 0.0557$$

We have enough SF to allow 3 SF in the answer.

COMMENT: It is sometimes easy to lose the sign in doing this kind of problem. By looking at the sign of the reaction as it is written, and seeing that this is the spontaneous direction of the reaction, meaning that we have to make the concentrations reverse this, we necessarily correct an error, if there is one. Also, the concentration of the solid phases is taken as one; a pure phase always has concentration one. If the cadmium, say, were in an amalgam with mercury, and was no longer a pure phase, but had a lower concentration, then this would have to be taken into account. In Chapter 12, we introduced Le Chatelier's Principle. This is not quite an example of that, but we do see that by having more Fe^{2+} than Cd^{2+} we would drive the reaction backward. The forward direction has $\mathcal{E}° > 0$, so the reaction must be driven backward to bring $\mathcal{E}°$ to 0.

3) Given the half reactions in Table 13-1, below:

> Part 1: Pick 2 of these half reactions that will produce a cell with the highest voltage
>
> *Estimate + Answer*: Obviously, if we want a high voltage cell, it would be good to start with a high voltage half-cell, so we choose the Cs half-cell to start; it has a large negative reduction potential, so it should be reversed. All of these are reductions, so the other electrode requires that we choose one of the other reactions as a reduction, and add the $\mathcal{E}°_{1/2}$ for that half cell. To get the highest cell voltage, we want to add the largest, in this case the Ag half cell. The cell reaction is therefore

$$Cs + Ag^+ \rightarrow Cs^+ + Ag \quad V = -(-2.923) + 0.799 = 3.722 \text{ V}$$

> Part 2: Find the equilibrium constant for this reaction at 300 K (consider T to have 3 SF)

$$\textit{Estimate}: K_{eq} = \exp(-\Delta G°/RT)$$

$$\mathbf{\Delta G°} = -z\,\mathcal{E}°\mathcal{F}$$

Table 13-1.

$Cd^{2+} + 2e \rightarrow Cd_{(s)}$	$\mathcal{E}°_{1/2} = -0.403$ V
$Fe^{2+} + 2e \rightarrow Fe$	$\mathcal{E}°_{1/2} = -0.440$ V
$Cs^+ + e \rightarrow Cs$	$\mathcal{E}°_{1/2} = -2.923$ V
$Ag^+ + e \rightarrow Ag$	$\mathcal{E}°_{1/2} = +0.799$ V

As written, this has $-z\,\mathcal{E}^\circ \gg 0$ so $\Delta G^\circ \ll 0$, and therefore $K_{eq} \ggg 0$. Even without the numbers, we could have guessed this, as cesium is a strong reducing agent, and silver is hardly a reducing agent at all. Obviously, the cesium is the one that "wants" to be oxidized.

Answer: $\Delta G^\circ = -(1 \times -3.722 \times 96530) = -359.3$ kJ

$$RT = 8.31 \times 300 = 2.493 \text{ kJ, so}$$

$$\Delta G^\circ/RT = -144 \text{ so } K_{eq} = 3.90 \times 10^{62}$$

Part 1 didn't specify the direction of the reaction, so we get this extremely large K_{eq}; we could have written the reaction in reverse, giving $K_{eq} = 2.57 \times 10^{-63}$. Of course, nothing physically has changed; you have all Cs^+ and Ag either way.

Part 3: Suppose 1 M NaCl is added to the Ag^+ ion half cell. The solubility product of AgCl (see Chapter 11, Solutions) is 1.6×10^{-10}. What difference does this make?

Estimate In this case, it probably doesn't matter much, because the equilibrium is lopsided by vastly more than 10 orders of magnitude. The concentration of Ag^+ drops from 1 M to 10^{-10} M. However, in a normal reaction where the equilibrium constant is not so great, it would shift the concentration ratio by 10 orders of magnitude, or the potential nearly 600 mV from what it would be at 1 M. Here the order of magnitude is way out of that range. As an exercise, consider the cell you could make from Cd/Cd^{2+} and Ag/Ag^+, where the potential is not so huge that ten orders of magnitude in concentration are not relevant.

COMMENT: What happens if T = 600 K? Essentially, what happens is that the problem ceases to make sense. These potentials refer to aqueous solution and water boils at 373 K. At high pressure, water boils at a higher temperature, but all sorts of things happen, including a requirement of pressure approaching 200 atm (water has a critical point of 647 K, where the liquid ceases to exist even at high pressure; the pressure to reach the critical point is 217 atm.) We would be so far from standard conditions that we could not use the normal table of reduction potentials. Such a problem is extremely problematic.

4) Concentration cell: We could make a cell for which $\mathcal{E}° = 0$ by using the same half reaction for anodic and cathodic reactions, but with different concentrations to drive the reaction. Simple consideration of what happens if you mix two solutions of different concentrations tells us that they will mix until the concentrations are the same, as the Second Law of Thermodynamics requires. We can separate the solutions so that the electron transfer has to go through an electrical circuit. In that case there is a driving force that is essentially the entropy increase above the entropy of the uniform solution; the entropy difference translates into a free energy difference, hence a voltage. The voltage difference derived from an entropy difference is enough to see that the direction of reaction must be that which causes the concentrations to change in the direction that will make them equal.

Consider the cell composed of Cs^+ (0.4221 M) vs. Cs^+ (0.03573 M) Find the potential at 298 K.

Estimate: Now we can be quite sure that $\mathcal{E}° = 0$. The potential is then 2.303 $RT/\mathcal{F}$ log $[Cs^+]_1/[Cs^+]_2$

The ratio of concentrations is greater than a factor of 10, so we expect the voltage to be greater than 59 mV.

Answer $V = 0.592$ $\log_{10}$ $[Cs^+]_1/[Cs^+]_2 = 63.5$ mV.

COMMENT: This cell has a smaller potential than most cells with different half reactions, in which $\mathcal{E}° \neq 0$. With the rate of change of $\mathcal{E}$ with temperature and the size of the cell (entropy is an extensive quantity), it would be possible to calculate the entropy difference that is responsible for the voltage, although we have not discussed this in this chapter. It has been proposed that energy could be derived from mixing fresh water coming from a river into salt water. So far, no one has proposed a practical way to do this. The general idea would be to have one electrode in the river, the other in the ocean. As an exercise, calculate the energy of mixing the flow of the Amazon River into the Atlantic Ocean, taking the ocean to be a NaCl solution, and the river to have 0.03 times the concentration of the ocean. Look up the flow of the river (it is dropping for several human-related reasons). This is a sort of engineering problem that probably has no practical solution.

16

Nuclear Chemistry

So far, all of the chemistry we have discussed has concerned the exchange of electrons, while the nucleus has been treated as a mass with positive charge, located at a point at the center of the electronic distribution, but otherwise inert. However, not all nuclei are stable. Some have too many protons for their number of neutrons, or too few. Neutrons, as isolated particles, have slightly more mass than protons do, and a neutron can decay into a proton and an electron; if there is a collection of neutrons, half would decay in this way in about 12 minutes. However, in a nucleus, neutrons are stable; the physics behind this involves the "strong force", which is otherwise not involved in ordinary chemistry, so we will just note it and move on. There are 81 atoms with at least one stable *isotope* (isotope: same number of protons, so same atom as far as chemistry is concerned), but the mass also depends on the number of neutrons. The word "isotope" means same place; all the isotopes, regardless of the number of neutrons, of a given atom, are defined by their number of protons, thus having the same atomic number, so they are in the same place in the Periodic Table. Some atoms, although not completely stable, live for a time comparable to the age of the earth, like uranium (^{238}U has a half life about equal to the age of the earth), and are found in nature. Nine elements, those with atomic number 84 through 92, are in this category. Two lighter elements, 43 (Technetium) and 61 (Promethium) have no stable isotopes, and exist, on earth at least, only in laboratories where they are made. A note on notation: $^{A}X_{N}$ means an atom with symbol X, atomic number (number of protons) N, and number of neutrons plus protons A;. A is not exactly the isotopic mass, as there are effects of the strong force, interactions with electrons, etc. However, A is close.

If there are too many protons, the nucleus will become more stable if a proton is transformed into a neutron. This can happen by having the nucleus capture an electron, or by emitting a *positron*, the anti-particle of the electron: it has the same mass and spin as the electron, but has positive instead of negative charge. Either way, with one less proton, the atom after this transition has been changed to the atom with atomic number one less, but close to the same mass. If there are too many neutrons for the number of protons, the nucleus can emit an electron, changing the atomic number to its original atomic number +1. In both cases additional particles and radiation (neutrinos, for example) may be involved. Typically the energies involved in these transformations may be hundreds of keV (keV = kilo electron volts, k = 1000 as usual). For heavier atoms, even MeV transitions occur (M = mega, 10^6). Considering that chemical energies are on the order of eV, tiny by comparison, nuclear reactions are very different.

One way in which we do have to consider the strong force (the force holding nucleons (nucleon means nuclear particle, proton or neutron, and stronger than the electromagnetic force) is by comparing it to the electrostatic force (the electromagnetic force for static electrical repulsion), as the protons repel each other. Eventually, as the number of the protons gets larger and larger, repulsion becomes so large that the nucleus becomes unstable. The heaviest atom with a stable isotope is bismuth (Bi), atomic number 83. All atoms with atomic number 84 or higher have no stable isotopes; however heavier atoms live long enough to have applications. The most notorious application is in nuclear weapons, for which two isotopes, $^{235}U_{92}$ and $^{239}Pu_{94}$, are used. There are two other modes of radioactive decay for these very heavy atoms. The common one is *alpha* decay, in which the heavy nucleus sheds the nucleus of a helium atom, with two protons plus two neutrons; as an alpha particle; this has very high energy. The heavy atom isotopic mass decreases by four while the atomic number decreases by two. With two fewer protons, the nucleus is less unstable, and there can be another alpha decay to bring the nucleus further toward the stable region, and so on until a stable nucleus is reached.

There is another mode of decay available to heavy nuclei with odd isotopic mass, like $^{235}U_{92}$ and $^{239}Pu_{94}$, in which the nucleus falls apart into two parts, each of which is a fairly large size nucleus, plus three to five neutrons, in the process called *fission*. The excess neutrons must come away separately, as the lighter nuclei have fewer neutrons per proton than the very heavy ones. These neutrons can produce additional fissions if they happen to hit another fissionable nucleus, which produces still more fissions, and so on, making a *chain reaction*. This means that, if there are so many such nuclei, or they are so tightly packed, that the probability that a single fission produces another fission approaches unity, you can have a chain reaction that will release much energy; if carefully controlled, this becomes a reactor, which can generate electricity. If uncontrolled, so that more and more fissions happen very rapidly, one has an atomic bomb. We will leave this topic, as it is not needed for introductory chemistry.

Nuclei have different degrees of instability, so they have different probabilities of decay per unit time. The decay is exponential; the rate of decay is proportional to $\exp(-t/\tau)$, where t = time, and τ is a time characteristic of each isotope. In other words, it is a first order reaction. Here there is no way to assume anything about orders of magnitude, as τ can range from small fractions of a second for some very unstable isotopes, to billions of years. For $^{238}U_{92}$, the more stable isotope of uranium, the time required for half to decay (mostly by alpha emission) is about 4.5×10^9 years, roughly the age of the earth. The less stable $^{235}U_{92}$ takes about 7×10^8 years for half to decay, still a long time. We consider the time for half the original quantity to decay, called the *half-life*, $t_{1/2}$, because that is convenient. $t_{1/2} = 0.693\ \tau$, where $0.693 = ln\ 2$. In other words, the number of nuclei decaying in a time $t_{1/2}$ is half the number of nuclei of that radioactive isotope that were originally present. This means that after $t_{1/2}$ half the original number of nuclei are still present. After another $t_{1/2}$ ¼ the original number of nuclei remain, after a third half life 1/8 remains, and so on. After n half lives, 2^{-n} of the original quantity of the isotope remains. For some isotopes, the toxicity and radioactivity is so great that even 10 half lives, leaving $\approx 10^{-3}$ of the original quantity, is not enough to say that the isotope is effectively gone. This is a reason that disposing of radioactive waste can be such a problem. Remaining plutonium, with a half-life of 24.000 years, may still be dangerous after a quarter million years.

SUMMARY: The energies of nuclear reactions are much greater than the energies of chemical reactions. Some nuclei have an unstable ratio of neutrons to protons, and there are several modes of reaction to remedy this imbalance. Very heavy nuclei are also unstable. Unstable nuclei can have lifetimes from billionths of a second to billions of years. We can calculate how much remains after a given time. In introductory chemistry, we do not go very deeply into nuclear chemistry.

Problems:

1) An isotope of atom X has a half-life of 23 minutes. A sample of this isotope is made in a cyclotron; after 1 hour, 32 minutes, the isotope is separated from interfering isotopes, and the isotope of interest is then measured and found to produce 10,000 gamma rays per minute. What was the rate of gamma radiation from this isotope when the sample was first taken from the cyclotron?

>   *Estimate*: Obviously there will be more decays at the early time than later, and there have been several half lives leading up to 1 hour, 32 minutes. We have 10^4 decays at the end of that period; it is reasonable to guess

that we should see around 10^5 decays at the beginning, since 3 half-lives means 8x, four 16x..

Answer: To get a complete answer, we need to find the number of half-lives from the beginning of the period to the time when the measurement was made. 1 hour 32 minutes = 92 minutes, or 4×23 minutes, so 4 half-lives. In this period the number of atoms will have decayed by a factor of $2^4 = 16$, so the initial rate of decay was 16×10^4 decays per minute.

2) You have a mixture of two isotopes, one of which, I1, has a half life three times longer than the other, I2. The initial concentration of I2 is ten times that of I1. How long, in terms of the shorter half life, will it take for the two isotopes to have equal concentration?

Estimate: We can almost do this by counting; after three short half-lives, you have almost as little. Set up a table of the concentrations at convenient times; the most convenient times are three short I1 half-lives, which are a single I1 half-lives life. We can tell when the two concentrations pass each other.

By 6 short half-life, I2 concentration has dropped below of that I1. Roughly 4 to 5 half-lives of I2 are needed.

Answer: Let the half-life of I1 $= 11$ (we want to avoid subscripts here, hence the unusual notation). We want to find the t at which $10 \times 2^{-t/11} = 2^{-t/311}$. Take logs, 0.693. Then, ln $(10) - t/11 \times$ ln $2 = - t/(3 \times I1) \times$ ln 2; solve for t: t = 4.98 11 half-lives.

Table 16-1.

I1 Half-lives	I1 concentration	I2 half-lives	I2 concentration
0	1	0	10
1	1/2	3	10/8>1/2
2	1/4	6	10/64<1/4

3)   In 100 kg of uranium, how many radioactive disintegrations per second do we get? The half -life is 4.5×10^9 years. Atomic mass = 238.

> *Estimate*: Here we have no obvious intuition. We need to know how many atoms we have, and the fraction that decay per second, which means converting from years to seconds. The atomic mass is close to ¼ kg/mole, so we have close to 400 moles, or about 2.5×10^{26} atoms. 4.5×10^9 year is roughly 10^{17} seconds, the time for half to decay. $10^{26}/10^{17}=10^9$. Therefore about $½ \times 2 \times 10^9 \approx 10^9$ atoms per second should disintegrate.
>
> *Answer*: $(100 \times 6.0 \times 10^{23})/0.238) \times (-0.693/1.42 \times 10^{17})$ $= 1.2 \times 10^9$ disintegrations per second.

COMMENT: We have effectively been assuming the 100 kg is as exact as we need. We still do not have more than 2 SF because the half life is known to only 2 SF. Moreover, we have ignored the minor isotope of uranium, ^{235}U, which is present to the extent of only about 1/140 the extent of the major isotope, but its half life is more like 1/6 that of the major isotope, so we expect a contribution of about 4%. This may not matter in the second SF (at most it drives it up from 1.2 to 1.3×10^9 disintegrations per second). Actually, the way the half-life is determined is by counting the number of disintegrations per second for a known mass of uranium, and working back to derive the half-life; it is not necessary to wait a time roughly as long as the age of the earth, 4.5×10^9 years, to measure the half-life. Of course, in a lab, one would not use 100 kg, but more like 100 mg. In 100 kg, most of the radiation from the decays would be trapped inside the mass of uranium, so the result would not measure the actual half-life in any case.

4)   Given 1.00 g of radium, half-life 1500 years, how long will it take until there is an even chance that there will be either 0 or 1 atom remaining.

Estimate: We need to know how many atoms we have, and then how fast they decay. There are about 1/228 moles, or around 3×10^{21} atoms. One half life cuts the number of atoms by a factor of 2; $2^{10} \approx 1000$, which is ten half-lives. $(1000)^7 \rightarrow 10^{21}$. In other words, after about $7 \times 10 = 70$ half lives, we are down to a just a few atoms.

Answer: To get to one atom when you started with 3×10^{21}, divide by 3×10^{21}. This is about $2^{71.25}$ half lives, or about 106,000 years.

COMMENT: Because we only have 2 SF, we can't really solve this problem as stated, The 71.25 half lives is not quite good enough to give us an answer to enough SF, and the 71.25 is itself too many SF. The question, with the given data, is not quite actually solvable; however, 71 half-lives, about 106.000 years, is a reasonable approximation. Because the time for the last atom to decay is a question of probability, and we can only guess how long it might take to go from two atoms to one, we probably need to say plus or minus at least a couple of hundred years.

5) Suppose there is a radioactive isotope that emits a beta ray of 700 keV. There are two milligrams of this isotope that has atomic mass 100, and the half-life of the isotope is 3.17 years. Find the total energy emitted in one week.

Estimate One year is 3.15×10^7 s, so 3.17 years is 1.00×10^8 s. In one week there is much less than one half life, so we can reasonably assume that the amount of the isotope is constant for the week. We simply have to multiply the number of disintegrations by 700 keV. To find the number of disintegrations, we observe first that we have 2×10^{-5} moles, so 1.2×10^{19} atoms, so 6×10^{18} atoms decay in 10^8 s, so 6×10^{10} s^{-1}. In 1 week, 6×10^5 s, we get $\approx 4 \times 10^{16}$ disintegrations. At 7×10^5 eV/disintegration, about 3×10^{22} eV. This is a problem which is unlikely to have any relevant chemical intuition so working it through with 1 SF helps only insofar as it is likely to avoid arithmetic errors.

Answer: In this case, we had to talk ourselves through the entire problem. All that remains is to plug in the actual values instead of the rounded-off values, to get 6.05×10^5 s week^{-1} This is only a 1% correction, and the rest of the numbers have been cooked to save time. The estimate is the answer to 2 SF. 1 eV = 1.6×10^{-19} J, and we just have to change units to Joules; 3×10^{22} eV $\times 1.6 \times 10^{-19}$ J eV^{-1} = 4.8 kJ.

Epilogue

We have been through the major topics of introductory chemistry, covering the background of each topic, but with the assumption that this was a supplement to a conventional text. Here, we emphasized numerical considerations. The point is that it is all too easy to lose track of what the "numbers" — the results of calculations — mean physically. Understanding the physical background, the meaning of the numbers, in turn allows you to rule out answers that are unreasonable, for example, an answer that suggests that there were tons of material in a 1 liter container — or similarly physically impossible results. Here we go through a very limited number of examples, but, hopefully, enough to suggest how to avoid this kind of mistake.

This kind of mistake often comes from depending on memorized equations, in which something is inverted, or a sign reversed, or a symbol substituted for another, or even a wrong button is pushed on a calculator. The equations get more difficult as one goes forward to more advanced material, and memorization eventually becomes impossible. If one pays attention to what the symbols mean physically, this almost always prevents ridiculous answers — it does not prevent you from making mistakes, even ridiculous results, but it does prevent you from accepting them, failing to realize that they must be wrong. If you have a diatomic gas that can partially dissociate, then starting with 0.1 atm of the diatomic gas, keeping everything fixed but the dissociation, you can get at most 0.2 atm of a combination of monomers and dimers; if you find 23 atm., something is wrong. Hopefully, going through this book will help you to catch such errors, so you will not let them stand. That is, this book should help you to approach the problem with a view to what is likely to be the range of reasonable answers, complete with units. "Reasonable" can mean both arithmetically and physically, with physical sense the most important. If you learn to think this way, you will be able to also work on more difficult problems, for which you may not even have a formula ready made. If you consider what you need to form a path to answering the problem, you have an excellent chance of finding that path. In doing science in general, it is necessary to have a physical feel for what you are doing; sometimes thinking about what makes sense, or is possible, leads you to a fruitful hypothesis.

In short, this book should help you to think physically and chemically about what you are doing, which in turn will help you to understand what is happening in your chemical system.

Appendix: An Extra Problem, with Some Extra Complications

1: An electrochemical cell is prepared from these two half reactions:

$$Cl_{2(g)} + 2e^- \rightarrow 2\,Cl^- \qquad \mathcal{E}^\circ_{1/2} = 1.36\text{ V}$$

$$Br_{2(l)} + 2e^- \rightarrow 2Br^- \qquad \mathcal{E}^\circ_{1/2} = 1.0873\text{ V}$$

Part 1: Find the voltage of a cell in which there is the reaction

$$Cl_{2(g)} + 2\,Br^-_{(aq)} \rightarrow 2Cl^-_{(aq)} + Br_{2(l)}$$

and then find the equilibrium constant at 20°C for the reaction. Assume the ions are in aqueous solution, and that there is 1 atm pressure of Cl_2 gas in equilibrium with the solution, and extra liquid Br_2 at the bottom of the cell; T = 298 K.

Estimate: First of all, this cell has a relatively small voltage, so the equilibrium constant is not going to be huge or tiny, at least in the sense that it's magnitude is inside the range 10^{-10} to 10^{10}; this counts as not gigantic. On the other hand, it is not small. The voltage will be positive if the reaction goes in the direction as written, so the equilibrium constant >1. Once we observe this much we can see that the voltage is 0.27 V (no need to estimate here; we only have 2 SF). The pure phases are not aqueous, but because they are the pure elements in the standard state, the dissolved Cl_2 and Br_2 are in equilibrium with the pure elements; both Cl_2 and Br_2 have appreciable solubility in water. Because the cell has an atmosphere of Cl_2 in its standard state, and there is liquid Br_2 at the bottom of the cell, these species are at the constant free energy of the pure elements, and this does not change as a consequence of the reaction.

Answer: ln $K_{eq} = \text{-}(\text{-}z\,\mathcal{E}^{\infty}\,\mathcal{F}/RT) = 2 \times 0.27 \times 96,530/8.31 \times 298 = 21.05$. Convert to base 10 and we get $K_{eq} = 10^{9.29}$. The reaction, however, is written for two electrons, or two atoms of Cl and Br.

Therefore,

$$K_{eq} = [Cl^-]^2/[Br^-]^2$$

We could have written the reaction as

$$\tfrac{1}{2}\,Cl_{2(g)} + Br^-_{(aq)} \rightarrow Cl^-_{(aq)} + \tfrac{1}{2}\,Br_{2(l)}$$

Either way, we have $[Cl^-]/[Br^-] = \sqrt{10^{9.29}} = 10^{4.65} = 4.4 \times 10^4$, where we round off to two SF at the end, because that is all the voltage difference allows. Although the question doesn't ask for it, we also get, as an intermediate result, the $\Delta G° = -z\,\mathcal{E}^{\infty}\,\mathcal{F} = -52.1$ kJ $= -52$ kJ for the 2 electron reaction (remember that free energy is an extensive property; $\Delta G° = -26$ kJ for the one electron reaction). We only have 2 SF because the voltage of the chlorine half reaction is given to only 0.01 V, and after subtraction only 2 SF remain; an extra SF is carried in the intermediate calculation to avoid additional round off error. There is no concentration for the pure phases, $Cl_{2(g)}$ and $Br_{2(l)}$, as is always the case for pure phases in equilibrium with their standard state. There is a lot more $[Cl^-]$ than $[Br^-]$ after the reaction. This is what to expect, of course, as chlorine is a stronger oxidizing agent than bromine. If we lost a sign, and found more Br^-, getting a reverse ratio, we would see at once that we had the reaction backward, just from our chemical intuition.

COMMENT: There is a lot that we did not have to include in this problem. What was the counter ion — the solution must be electrically neutral. The concentration? Does it matter? Yes, but this is a correction to the problem as stated, and the correction cannot be calculated with the tools that we have used in first year chemistry; the corrections deal with the interactions of the ions. We did briefly discuss the van't Hoff factor in the section on colligative properties, noting that it would not be exactly the number of ions, but we did not go any further towards getting the final value. Similar interactions are important here. to the extent that we take these into account at all. In the next part of the problem, although we still cannot get this correction, at least the counterion and concentration are relevant.

> Part 2: The solubilities of Cl_2 and Br_2 are not negligible. The approximate solubilities are: Cl_2 0.116 m at 20°C and 1 atm pressure, Br_2 0.224 m at 20°C (caution: these are a little off, so don't use them without checking — they are within about 10% of the actual values, but adjusted slightly for convenience). Can we still get the same results as in Part 1? We can, on condition that the solutions are in equilibrium with the pure phases, which are in their standard states, and the potential is defined accordingly; at the beginning, we were told that in the cell above the solution, there is Cl_2 vapor at one atmosphere, and at the bottom of the cell there is liquid Br_2. We should write the equilibrium constant as
>
> $$K_{eq} = [Cl^-][Br_2]/[Br^-][Cl_2]$$

But we drop the pure elements in their standard states. Then the free energy of the molecules in the solution must be the same as that of the pure phases, or else they would be out of equilibrium. So here is a question: how much difference would it make if the pressure of the Cl_2 above the solution, and in equilibrium with it, were dropped to 0.5 atm? The solution reaches equilibrium with this new condition. Find the difference in cell potential and in ΔG.

Estimate: To begin with, we don't expect a huge difference. Also, there is less Cl_2 dissolved in solution as we must assume that the Cl_2 concentration is at least approximately proportional to the Cl_2 pressure above the solution. Le Chatelier's principle tells us that the reaction would have to restore some Cl_2 so it would not go as far forward as when there was more Cl_2. Using the assumption that the quantity of Cl_2 in solution is proportional to the pressure of the gas above the solution (if the assumption is off a little, since this will be a correction term, and we have only 2 SF to begin with, this should be safe enough). This time we have to include the $p(Cl_2)$ in the reaction, because we are no longer in the standard state, and the free energy changes.

Answer: If the solution $[Cl_2]$ is proportional to $p(Cl_2)$, the change in free energy is $-RT\ ln\ (1/2) = +0.693\ RT = 1.7$ kJ, making the overall $\Delta G = 50.4$ kJ. In calculating the concentrations, this has to be taken into account. The equilibrium constant, as defined with standard states, does not change. The cell potential would be smaller (the reaction does not go forward as far) by $\Delta E = RT/\mathcal{F}\ ln\ 2 = 0.016$ V. Rounding $\Delta G = -50.4$ kJ to -50 kJ may not be a good idea, as the ΔG appears in an exponent.

Part 3: Go back to the conditions of part 1 except that the Cl_2 pressure above the cell is unknown but take into account the solubilities given in Part 2 for Cl_2 and Br_2. The counter ion is sodium, and the concentration of the NaCl and NaBr was 0.100 m each. The freezing point of the solution is found to be $-1.30°$. Can you find the concentration of Cl_2 in solution? Pretend that the solubilities of the molecules are the same at the freezing point as at 25°C (they change a fair amount in reality). The cryoscopic constant for water is 1.86°C (the freezing point depression per molal concentration).

Estimate: A quick glance at what we have in solution adds up to something around half molal. Add all the molalities other than Cl_2, find the freezing point depression that that much would produce, and see how much more the Cl_2 would add to get to 1.30°C; then get the amount of Cl_2 needed to produce this.

Answer: We have to add the molalities of all the species in solution. We have the following species: Cl^-, Br^-, Na^+, Cl_2, Br_2. In Part 2, we are given the molality of Br_2 as 0.116 m. There is excess liquid Br_2 in contact with the solution, so it is safe to assume the solution is saturated. We must find the concentration of Cl_2 in the solution, as the concentration depends on the partial pressure of the gas in equilibrium with the solution. This does not mean the solution is saturated. In the reaction, the molalities in the solution of the pure elements do not change, as they remain in equilibrium with the elements in their standard state. The ions may switch to or from the elemental form, but the total molality of $Cl^- + Br^-$ does not change, so the total of negative ions remains 0.200 m, as needed to match the Na^+ concentration, which does not change. This makes 0.400 m; add the 0.224 m Br_2, for 0.624 m, leading to a freezing point depression of 0.624 m $\times$ 1.86°C m^{-1} = 1.16°C. Then the additional drop in freezing point from the Cl_2 is 0.140°C, which requires the Cl_2 concentration to be 0.14/1.86 = 0.075 m to get the full 1.30°C. Since 0.075 < 0.116, the solution is less than saturated; if it had required a concentration greater than saturation something would be wrong. In this case, we can actually save a third SF, as it is unaffected (to the approximation we are using) by the reaction, where the bromine half reaction is responsible for the loss of 1SF. What is more, we need the third figure to get a meaningful answer for the molality of the Cl_2. If we go to a higher approximation, in which the ion – ion interactions were taken into account, we would find a slightly different answer; the van't Hoff factor would become complicated.

COMMENT 1: The numbers in this problem have been cooked, so don't take them to be exactly correct; the point of the problem is to suggest how to look at the problem.

COMMENT 2: Suppose we added $AgNO_3$ to the solution, but less than enough to precipitate all the chloride and bromide. What would change? The silver ion precipitates with both chloride and bromide. If there is less than enough $AgNO_3$ for the anions, the AgBr, which is less soluble than the AgCl, will precipitate first. Then the potential changes, as the reaction proceeds further to the right (LeChatelier, again). The total concentration will not change for either cations or anions, as the nitrate replaces the anions that precipitate. The silver precipitates, if more is added, until all the Cl^- and Br^- have precipitated, but the sodium remains, so the concentration of cations is also unchanged. While the K_{sp} of AgCl and AgBr are finite, the amount that remains in solution is literally negligible compared to the major ions — neglect it. If excess $AgNO_3$ is added, the cell as written originally is essentially destroyed, as the Br^- and Cl^- are removed. There may be a reaction in which the Br_2 and Cl_2 that remain in solution are involved, but now we are talking about a different cell.

FINAL COMMENT: There are a couple of subjects we could not deal with in this Appendix Problem. Rates could not be included, because here the reaction occurs on a very short time scale once the reactants come in contact, so the rate is determined by the rate of diffusion of the reactants, a topic we could not cover in an introductory course. A catalyst could not help here. This set of reactions does not include protons, so pH is omitted. Because there are redox reactions that do involve protons, a pH dependent problem could exist, but it would be more complicated. However, there is enough in this problem to cover several parts of the course, as do the fundamental ideas of making sure that answers make sense, something that we could look at from several points of view here. In this problem, we also see that several topics, although meriting separate chapters, really interact; we do not get to ignore one topic when a problem is labeled as related to a seemingly different topic. At this point you should be able to start making up your own problems. Can you find a redox reaction that involves hydrogen ions? How about involving a reaction that is slower, not diffusion limited, but rate limited by a slow step. If you can't find data make up reactions with numbers that allow a calculation, even for molecules labeled X, Y, and Z, or anything else. Combine these to make a problem that involves pH, reaction rate, and an electrochemical cell. Perhaps you can think of other problems. If you understand physically and chemically what is happening in a cell, you should be able to combine ideas so that you can calculate what happens in a wide variety of chemical situations.

As long as we keep in mind that there is a way to make sure that we get reasonable answers, we will avoid the most common mistakes in introductory chemistry.

Index

absolute zero, 25, 152

absorption coefficient, 131

absorption of light, 70

acid and base, 188
   Arrhenius, Svante, 182
   Bronsted, Nikolaus, 182
   conjugate acid, 184
   conjugate base, 184
   indicator, 186
   Lewis, Gilbert N., 183
   Lowry, Thomas Martin, 182
   pH, 188
   titration endpoint, 187
   weak acids, 183

acid and base, definition, 182

acid rain, 63, 64

acids
   acid constant, 183
   conjugate acid, 183

activation energy, 197, 198, 203, 211, 212

activity of the solute, 173

adiabatic flame temperature, 166

air pollution, 141, 209

alcohol, 143

alkali cations, 173

alpha decay, 228

Amazon River, 225

amorphous, 111

Ampere, 25, 213, 214

amu scale, definition, 34

angular momentum, 73

anode, 216

aqueous solutions, 113, 181

aromatic, 93, 95

atmospheres, 42

atmospheric inversion, 141

atomic mass unit, 33

atomic number, 34, 228

atomic radius, 103

Auger radiation, 90

Avogadro, Amadeo, 26, 45, 61

Avogadro's number, 26, 37

Balmer series, 88

bar, 25, 42

barometer, 42

barrier, 197, 203

Batteries, 213

Beer's Law, 122

biology, 199

bismuth, 228

Bohr, Neils, 71, 87
boiling point, 117, 120
    ebullioscopic constant, 121
boiling point elevation, 115, 119
Boltzmann's constant, 26
bond energies, 136, 155
bounds, upper and lower, 6
Bronsted definition, 183
buffer solutions, 185
    buffer capacity, 185

calculus, 53, 66, 67, 92
Cannizzaro, Stanislao, 26, 45
carboxylic acid, 143
catalysis, 199
catalyst, 197, 203, 243
cathode, 216
cell voltage, 222
chain reaction, 229
chemical bond, 71, 198
    broken bond, 195
chemical reaction, 55, 151
Chlorophyll, 75
Clausius, Rudolf, 150
closed system, 133
colligative properties, 115, 119, 123
combustion, 166
concentration, 113, 114, 116, 123, 141, 142
Concentration cell, 224
concentration differences, 218
concentration ratio, 219
conduction band, 96
conductivity, 123
conjugated double bonds, 82
conjugate quantities, 78

conjugate variables, 80
conservation of charge, 189
conservation of charge and mass, 187
conversion factor, 20
Coulomb, 213, 214
Coulomb potential, 80
covalent solids, 97
critical point, 120, 204, 224
cryoscopic constant, 241
crystal
    cubic crystal, 107
crystal, hexagonal, 98
crystalline, 111

Dalton, John, 45
Dalton's Law of Partial Pressures, 44
Davisson and Germer, 81
de Broglie wavelength, 87
Debye, Peter, 70, 173
degree of dissociation, 194
degrees of freedom, 79
density, 26, 104, 105, 107, 109, 110, 118
dependence of rate on concentration of reactants, 200
depression, 172
desalinate water, 115
deuterium, 36
diamond, 97
diffraction grating, 98
diffusion, 45, 46
Dimensionless variables, 26
dimerization, 133
disintegrations, 233
dissociate, 173

dissociation, extent of, 138
DNA, 76
Donora, PA, 142
double bonds, 75
dr Broglie, Louis, 81

effuses, 45
Eigen ion, 182
Einstein, Albert, 70, 85
electric current, 213
electric field around a charge, 181
electrochemical cells, 213, 215
electrochemical reactions, 220
electrodes, 220
electromotive force, 215
electron, 215, 218
electron cloud, 71
electron density, 71
electronic levels, 82
electron orbitals, 73
electron volt, 78
elementary ideas, 3
emitted electron, 86
endothermic, 154, 198
energy, 24, 72, 150, 218
    kinetic energy, 24
energy of chemical bonds, 82
enthalpy, 149–151, 153
enthalpy of formation, 152
entropy, 115, 149, 151, 153, 155, 172, 225
    absolute entropy, 152
environmental problems, 30
enzyme, 199, 203
equilibria, 173, 215

equilibrium, 133, 137, 138, 149, 169, 209, 239
equilibrium constant, 133–135, 143, 153, 170, 182, 184, 223
ester, 144
esterification, 209
eV, 214
excess reagent, 65
excited state, 82
exothermic, 154, 198
expansions, 11
extensive property, 148
external circuit, 213

Faraday, 213
fission, 229
force, 24
fortnight, 19
free energy, 115–118, 133, 134, 137, 147, 149, 151, 195, 197, 213, 238
    maximum free energy, 195
free energy of activation, 196
free energy of formation, 135, 136
freezing point, 117, 120, 130, 172
freezing point depression, 115, 119, 121
frequency, 69, 79
furlongs, 20

gas phase, 141
glasses, 97, 98, 104
ground state, 82, 152

half life, 227, 229, 230, 232, 233
half-life of radium, 177
half reaction, 219, 220
Hamilton, William Rowan, 70
harmonic oscillator, 76, 77, 92
heat, 148, 149
heat capacity, 150, 152, 153, 168
heat capacity, molar, 152
heat capacity of solids, 70, 71
Heisenberg Uncertainty Principle, 78, 80
hemoglobin, 75
Hertz, 82
Hess's Law, 154
Hooke's Law, 77, 92
Huckel, Ernst, 173
Huckel's rule, 95
hydration, 171, 174
hydrogen atom, 77
hydrogen atom spectrum, 87
hydrogen bonds, 76
hydrolysis, 183, 185, 188–190
    hydrolysis constant, 184
hydrolyze, 143, 187
hydronium, 181, 216
hydroxide, 216
hypothesis, 236

ideal gas, 41, 42, 44, 46
ideal gas law, 41, 47
ideal solution, 122
infrared, 74, 80
infrared region, 78
Inner shell fluorescence, 90
integrating factor, 151

Intensive and extensive properties, 26
intensive property, 148
interaction among ions, 114
interface, 140
interference fringes, 110
intermolecular forces, 97
interplanetary space, 46
intuition, 5, 7
ionic crystals, 97
Ion interactions, 172
ionization, 193
ionization of water, 182
IR spectroscopy, 72
isotope, 34–36, 38, 39, 227, 231, 233

Joule, 25

Kelvin, 25, 42, 142, 150
Kendrew, John, 98
kinetic energy, 41

Lavoisier, 60
Lavoisier, Antoine, 61
Laws of Thermodynamics, 150
LeChatelier, 242
Le Chatelier's principle, 170, 171, 222, 240
light frequency, 72
limiting reagent, 57, 59, 65
Limits, 5
limits of the accuracy of the data, 3
limit, upper or lower, 2
linear equation, 9

liquid phase, 155
liquid-vapor transition, 120

machine, 150
mass conservation, 189
maximum
   barrier, 195
mechanism, 207
membrane, 118
Mendele'ev, Dmitri, 45
metals, 97
molality, 113, 120
molar concentrations, 134
molar heat of formation, 158
molarity, 113, 132
molar volume, 46
mole, 37, 46, 55
Molecularity, 202
Molecular mass, 35
molecular rotations, 93
molecular vibration, 196
mole fraction, 44, 114, 129
momentum, 82

neutrino, 228
neutron, 33, 34, 227
Newton, 25
Newton's Second Law, 25
node, 71, 73
non-covalent bonds, 76
non-linear expression, 11
normal modes, 78, 92
nuclear reactions, 230
nuclear weapons, 228
nucleon, 228
nucleus, 33, 227

order, 207, 208
orders of magnitude, 5
Osmosis, 116
osmotic pressure, 115, 118, 119, 129, 130, 172
oxidation, 216
oxidized, 215
oxidizing agent, 219

π electrons, 93
parabolic potential, 77, 92, 97
partial pressure, 44, 139, 164
particle in a box, 76, 77, 93–96
partitioned, 151
Pascals, 25, 42
path dependent, 148, 149
path independent, 148
Periodic Table, 45, 227
perturbation expansion, 12
Perutz, Max, 98
pH, 185, 189, 243
phase, 113
phase boundaries, 120
phase boundary, 117
phase change, 44
phases, 97
photelectric effect, 85
physical background, 235
physical sense, 236
Planck, Max, 69
Planck's constant, 69, 70, 74
polarized, 94
polar molecule, 181
polymers, 99
positron, 228
potential, 76

precision, 114
pressure, 24, 25, 46, 118, 138
protein, 33, 76, 98, 199, 214
proton, 33
pure phases, 239

q_{rev}, 153
quantum, 69
quantum mechanical, 91
quantum mechanical reason, 93
quantum mechanics, 80, 181

radiation, 90, 228
radioactive decay, 228
radioactive disintegration, 232
radioactivity, 36
radium, 232
Raoult's Law, 121, 122
rate, 196, 199, 242
rate constant, 203, 208, 211
rate of reaction, 66
reactant concentrations, 209
reaction, 196
reaction coordinate, 197
reaction equilibrium, 182
reaction order, 200, 201
    first order, 200
    second order, 200
reaction path, 197
    intermediates, 200
reaction quotient, 171
reaction rate, 195, 203
    concentration dependence, 203
    fast and slow Steps, 202
    rate constant, 201
    reaction, 210

second order reaction, 201
redox, 215
redox reactions
    half reaction, 217
    oxidation, 213
    reduction potential, 217, 219
reduced, 215
reduction, 213, 216
reduction potential, 218, 219
reverse osmosis, 115, 118
reverse reaction, 209
reversible, 148, 149
reversible path, 152
RNA, 76
roots, other, 10
rotate, 75
rotation, 75, 76, 79
Rydberg constant, 87, 89
Rydberg levels, 77, 80

saturated solution, 169, 177
saturation, 170
Schrodinger equation, 87
Second Law of
    Thermodynamics, 224
second order, 207
shells, 73
significant figure, 5, 12
    Addition and Subtraction, 14
    Logarithms, 14
    Many step calculations, 15
    Multiplying by an exact
        number, 13
    Multiplying or dividing, 13
slide rules, 12
solid phases, 120

solubility, 238
solubility of ions, 173
solubility product, 223
solute, 113, 116, 117, 169, 170
solvent, 113, 116
spectra of atoms, 71
spectrophotometry, 122
spectrum, 70, 80
speed of light, 20
square roots, 9
standard states, 135, 152, 238, 239
standard temperature and pressure, 27, 42
state function, 149
stoichiometric coefficients, 57, 133, 154
stoichiometric equation, 61
stoichiometric problem, 58
strong force, 33–35, 227
successive approximations, 27
superconductors, 97
surface, 99
surroundings, 147, 148
symmetry, 100, 104
system, 147, 148
    closed system, 148
    isolated system, 148
    open system, 148

temperature, 25, 42, 46, 198, 211
temperature dependence, 172
temperature, high, 70, 138
thermal energy, 26, 70, 76, 137, 196
thermal motion, 81

thermodynamics
    First Law, 150, 151
    Second Law, 150, 151
    Third Law, 152
third order reaction, 207
titrations, 186
toxicity, 141
translation, 79
trapezoidal approximation to the integral, 67
Trouton's Rule, 154

ultraviolet, 74, 75, 80
ultraviolet catastrophe, 69
unit cell, 98–100, 102, 104, 105, 108, 110
    atomic packing fraction, 103, 104
    body centered cell, 100
    body diagonal, 102
    face centered cubic, 100
    face diagonal, 102
    unit cell volume, 103
universe, 148, 153

van der Waals, 47
van der Waals equation, 27, 43, 44, 46, 58
van't Hoff, 120
van't Hoff constant, 121
van't Hoff factor, 119, 130, 239
vapor, 155
vapor phase, 140
vapor pressure, 139, 140
velocity of light, 69, 72
vibrate, 75

vibrational transitions, 82
Virial expansion, 44
visible, 80
Volt, 213
voltage, 214, 215, 218
volume, 46

water, 98, 216
wavelength, 69, 72, 75, 76, 82

wave number, 72, 78
work, 149
work function, 85, 86

X-ray, 74, 90, 98

Zundel ion, 182

www.ingramcontent.com/pod-product-compliance
Lightning Source LLC
Chambersburg PA
CBHW072212150726
48002CB00005B/1776